sun Reflections

Sun Reflections

Images for the New Solar Age

By John N. Cole

Accompanying Material by Sara Ebenreck

Rodale Press ❧ Emmaus, Pennsylvania

Design by John F. Carafoli
Layout by Jerry O'Brien

Printed in the United States of America on recycled paper, containing a high percentage of de-inked fiber.

Library of Congress Cataloging in Publication Data

Cole, John.
    Sun reflections.

    Bibliography: p.
    1. Sun. 2. Sun (in religion, folk-lore,
etc.) 3. Solar energy. I. Title.
QB521.C63   1981   523.7   80-20997
ISBN 0–87857–318–6 hardcover
ISBN 0–87857–317–8 paperback

2 4 6 8 10 9 7 5 3 1 hardcover
2 4 6 8 10 9 7 5 3 1 paperback

Table of Contents

Acknowledgments   vi

Introduction   1

Chapter 1   SUN FACT   7

Chapter 2   SUN FANCY   43

Chapter 3   SUN NATURAL   75

Chapter 4   SUN PERSONAL   113

Chapter 5   SUN LIVING   145

Chapter 6   SUN TECH   189

Chapter 7   SUN FUTURE   217

Bibliography   236

Credits   240

Index   248

Acknowledgments

Writing a "book about the sun" is a concept that began with the creative thinking of Bob Rodale. None of us was quite sure of the book's precise dimensions when we began discussions about it, but each of us agreed with Bob's foundation premise: if contemporary citizens could be reminded of the cultural, mythical, and spiritual importance of the sun throughout human history, then we might move with more assurance and understanding toward a solar society. It sounds like a rather obvious notion now, but it is not, nor was it when Bob Rodale presented it. He is the father of the idea that makes this book possible.

Editors Carol Stoner and Marcy Posner worked hard to be certain that my writing never strayed too far from the Bob Rodale premise. At times, their job was difficult, yet they kept the book within bounds and in focus.

And Sara Ebenreck not only contributed the quoted material that adds so much, but provided me with a great deal of source information and the benefit of her considerable insight as well.

These four people, along with my wife Jean, who typed my manuscript and has made all my books possible, are due your gratitude; they already have mine.

John N. Cole

INTRODUCTION

The past is but the beginning of a beginning, and all that is and has been is but the twilight of the dawn.

H. G. WELLS

INTRODUCTION

In the 400 years since Copernicus accurately declared the sun, and not our earth, to be the center of our particular solar system, we have tried to displace that fact of physics. With accelerating determination—much as a meteor accelerates as it is pulled to its destruction by ever-increasing gravitational forces—we have used technology to become more and more earth centered. During the past century, the distance between us and the sun has become a widening chasm. Combining energy the sun stored beneath the earth's skin millions of years ago with a stunning array of technological inventions, we have done our best during recent decades to create environments that are closed to the sun, independent of natural forces.

Our windowless buildings are mechanically heated and cooled during winter and summer, illuminated fluorescently, decorated with plastic rubber plants, and saturated with a continuous flow of bland music. Most Americans now live and work much of their lives within these spaces; for them, weather reports are mere minutiae. What does it matter from which quarter the wind swings, how gentle the rain, what moment the sun rises and sets? Such wonders are of little or no concern in a culture which has been persuaded that the mastery of science is also the mastery of Nature.

Were it not for the current concern over the recognition that fossil fuels are indeed finite, and thus limited, we would have discovered few reasons to re-establish our relationship with the sun. Even now, as the debate over solar energy becomes a discussion of national significance, the sun is not viewed with any vestige of past reverence, but is seen

as an "energy source," a producer of so many energy units per square foot of the earth's surface. If only those units could be captured and stored—via yet another technological breakthrough—so the reasoning goes, we could acquire a limitless energy supply to accelerate even further our rush from the natural.

Indeed, some of us have become so removed that we are in the process of creating our own suns, here on earth. It is nuclear energy, in its fission and fusion forms, that is at the opposite pole of the energy debates. Those who see the sun as an energy producer are on one side of the table; those who would split and disintegrate the atom in an earthbound replica of the sun's core are on the other. It is the ultimate indicator of our hurtling journey away from the sun that some of us now propose, in fact favor, creating our own suns here on the planet. It is this process that now has most momentum, even though the risks are acknowledged to be meteoric: fusion, out of control, will leave only a comet's tail.

During the millions of years that life has existed on this earth, the sun has seldom been seen solely in its two current modes: one as a fusion phenomenon to be duplicated, the other, an energy source to be captured and exploited. It has, instead, been viewed through a third dimension—a dimension of wonder, of reverence, of mysticism, magic, worship, gratitude, artistic inspiration, and an awareness of essential mystery. These are the qualities so vividly absent from today's perceptions. Yes, we have astronomers, we have physicists, we have more knowledge of the facts of the sun than ever before. But in a paradox of history, as that physical knowledge has increased, the sun's metaphysical significance has decreased until the persona of our star is eclipsed by the data of its vital statistics.

Just as we shall never be governed best by leaders chosen only for their height, weight, features, age, and sex, so we shall not establish a good and true relationship with the sun until we re-explore its metaphysics, re-read its fables, re-acquaint ourselves with its mysteries and re-invest it with the full measure of meaning for humanity that the

sun has acquired since it first generated the environment for life on this earth millions of years ago.

The wonder is that a force as pervasive as the sun's could have been as minimized as it has been since the start of the Industrial Age. One has only to consider the fundamental universality of the sun's relationship with this planet to comprehend the totality of its effect on our lives. The sun provides heat and light; without it, we would indeed freeze in the dark. It is the parent of the wind, the progenitor of clouds, the creator of rain, the sprouter of seeds. Our seasons are orchestrated by the sun's arc in the heavens; our sunrises and sunsets are witness to its daily passage—a journey that moves a perpetual dawn around the globe, inspiring bird songs in a rippling wave from meridian to meridian so that there is never a moment when this earth is without natural music.

The sun tames glaciers, is the alchemist that draws life-giving oxygen from the plants of land and sea, and is the massive gravitational force that holds each celestial sphere of our galaxy in its proper orbit. Without the sun, humanity would have long since been lost in space.

Echoes of that knowledge reverberate within us, unconsciously, whenever we feel the warmth of the sun in our face, whenever we awaken to a bright dawn. We say we "feel great" in the clear sunshine of a new spring, or on an azure September afternoon; what we are feeling is the sun's own reassurance that the natural order of things is in place. We say "it came to me in a flash" when we have an idea, a creative thought; the imagery is one of interior illumination, of a light within the mind, of a small sunbeam flashing within the cellular galaxies of our own internal universe. The parallels are no accident; the electromagnetic energy that allows brain cell communication is a duplicate of the sun's radiations. Although this earth captures but a fraction of that energy (the rest is bestowed munificently on our sister planets and the vast void of space), it is, on balance, more than enough, not only for our survival, but for our flourishing.

We can comprehend and have proven the mathematics of the

solar energy equation. What we have been unable to do in recent years is to acknowledge the sun as our primary natural presence—metaphysically as well as physically. If we had done that, we would not view the future with anxiety but with the certainty that it could be lived to the fullest, in harmony with the natural.

It is the purpose of this volume to help you see the sun for its wonder as well as its warmth; for its history as well as its hydrogen; for its relationship to the individual as well as its spatial relation to earth. Only by re-investing the sun with these dimensions can we hope to establish it as a benign and primary energy source. The sun has, after all, been around for some 4,000 million years, and (we are informed by astronomers) should be around for at least 5,000 million more. It behooves us—energy crisis or no energy crisis—to get to know it in all its fascinating aspects, not just its vital statistics.

It is a comment on the technological imbalances of this, our generation, that we could have been led so far from the sun. Our solar balance needs restoring; this book is a beginning.

**Chapter 1
SUN FACT**

The night has a thousand eyes,
And the day but one;
    Yet the light of the bright world dies
    With the dying sun.

                    FRANCIS WILLIAM BOURDILLON

1   *SUN FACT*

Writing more than half-a-century before the birth of Christ, the Roman poet Lucretius stated: "Therefore, things can not be reduced to nothing." (The Nature of the Universe).

Eighteen centuries later, the Second Law of Thermodynamics held that any functioning system or organization produces predictable amounts of entropy—the word for an abstraction that includes such qualities as uselessness, noise, confusion, decay, and other elements in the process that eventually, even for a planet or a star, lead to death.

But in 1977, Ilya Prigogine, a Belgian chemist, won a Nobel Prize for experiments which challenged the validity of the Second Law. Observing a series of chemical reactions, Dr. Prigogine discovered that cells of greater complexity form in certain fluids when heat is passed through them. In other words, there is evidence of an inexorable force pushing life and humanity to further evolution and complexity, not decay and death.

According to the Second Law of Thermodynamics, the sun, our star, has some 5,000 million years left to live. According to Lucretius and Dr. Prigogine, there is reasonable room for doubt. One theory directs humanity toward a brighter future, another to an ever darker world. Both predictions, however, are theory, not fact. Theories, unlike a brick dropped from a third-story window, are not irrevocable. They change with time, with discovery, with physics and metaphysics.

There is a moral in the volatility of theories. Like the needle of a compass, it points to a philosophical principle which will act as a guide

In the beginning, after the word, was the sun. Through all the ages and stages of life on this planet, up the long dark ladder of time from cellular void to genetic variety, from ooze to intelligence, shining across all the pages of geologic and recorded history, presiding over each day and bringing on each night, forcing the growth of green plants, controlling the climate, creating fire and destroying ice, giving light and warmth, oneness and continuity and beauty and meaning—was the sun.
J. G. Mitchell, Introduction to Dennis Stock, *Brother Sun*

[*Continued on page 15*]

9

There is geometry in the humming of the strings. There is music in the spacings of the spheres.
Pythagoras, fifth century B.C.

Driving due west on a superhighway at 70 m.p.h. in late afternoon with the sun in our eyes we explain to the children that the earth's surface—like a phonograph record's outer rim—revolves faster at the equator than near the poles and that at mid-U.S.A. the surface is revolving eastward, away from the sun, at 860 m.p.h., which means that we are running a losing race to keep the sun in sight and the children excitedly see, and so do you, the extraordinary speed of eastward revolution of the earth speeding by the car as we get swept back eastward at 790 m.p.h. despite our 70 m.p.h. westward rate. Thenceforward the children "see" the true Earth-Sun relationship realistically.
Buckminster Fuller, *Utopia or Oblivion: The Prospects for Humanity*

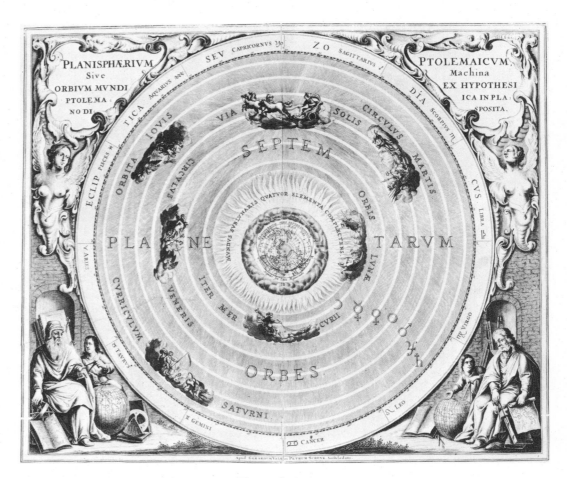

Ptolemy's conception of the universe. The earth is in the center with the sun and known planets revolving about it.

Copernican conception of the universe. Here the sun is in the center with the earth and other planets revolving around it.

The meaning of this seemingly endless music, to me, is that not only does the atom sing but it sings in time with the stars and with its whole surround in space-time, resonating in significant harmonies with other atoms and with all sorts of energies, known and unknown.
Guy Murchie, *The Music of the Spheres*

During the fourth century A.D. *the Roman architect Faventinus advised wealthy owners of the large villas he designed to put dark sand, soot, and ashes in a shallow pit under their dining tables. Faventinus had learned that as the sun shone on the heat-absorbent material during the day, enough heat could be stored to keep the floors warm during the late evening hours when the noble Romans and their guests gathered for their sumptuous meals. Faventinus assured his clients that the system would work and ". . . will please your servants, even those who go barefoot . . ."*

The leading European theory of the solar system in the fifteenth century was based on the second-century theory of the Alexandrian, Ptolemy. According to this theory, the earth was fixed in the center of a structured cosmos with the other known planets and the sun whirling in circles about it.

With a revolutionary imaginative leap, Copernicus saw that the motions of the planets could be explained more simply if looked at from the viewpoint of the sun, not the earth. "The sun rules the family of stars," he wrote, "and the planets including earth rotate about it in a harmonious system."

The work of this solar giant was not taken seriously by the rest of the world until half a century later when it inspired Galileo and Kepler. Today too, we might make further leaps if we looked from the viewpoint of the sun.

NICOLAUS NICOLAI COPERNICUS THORUNENSIS ARTIUM ET MEDICINÆ IN UNIVERSITATE CRACOVIENSI DOCTOR. CANONICUS VARMIENSIS DISCIPLINÆ ASTRONOMICÆ IN EUROPA INSTAURATOR NATUS ANNO M CCCCLXXIII DIE FEBR: XIX. OBIIT AN: MDXLIII. DIE MAII XXIIII.

Copernicus (1473–1543).

STRUCTURE OF THE UNIVERSE

GALAXY OF STARS   ## SOLAR SYSTEM

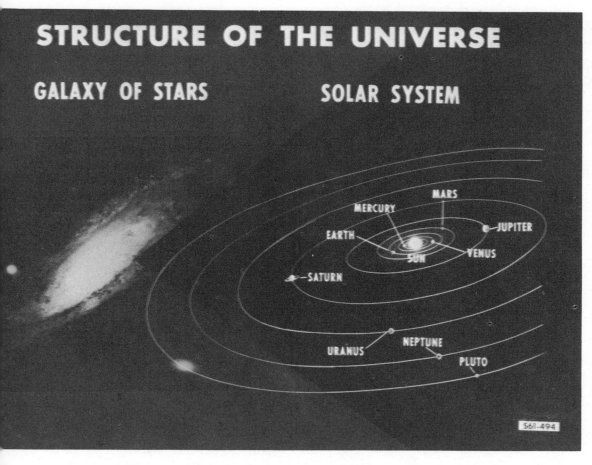

S61-494

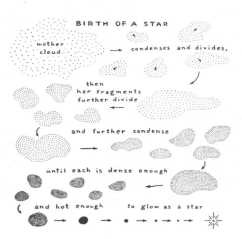

BIRTH OF A STAR

mother cloud → condenses and divides,

then her fragments further divide ←

and further condense →

until each is dense enough ←

and hot enough   to glow as a star

The 100 billion stars in our galaxy are arranged in the form of a flattened spiral. The sun and its family lie well away from the center, about one-third of the way toward the edge. The spiral is so large that a flash of light from one edge would take 100,000 years to reach the other side. The sun has its own path in the galaxy, making an orbit every 200 million years. And the galaxy itself revolves at the rate of about 200 miles per second.

13

Galileo's development of the telescope from the status of a curious toy to a scientific instrument opened the way for a new descriptive science of the sun. First to observe sunspots with clarity, Galileo saw they were indeed part of the sun and not floating matter in space. Looking at Venus with his telescope, Galileo saw this planet had phases like the moon, showing it was illumined by the sun and thus that it revolved around the sun in the way Copernicus had predicted. The Church, trying to shore up its defense of the Ptolemaic theory, tried Galileo and forbade the teaching of his observations.

Galileo (1564–1642).

for much of what follows herein. Stated bluntly, in so many words, that principle, the "first law" of this book might read: "What humanity has learned about the universe is but a pebble on a mountain compared to all that is yet to be learned."

While that first law of this volume is a bit of illustrative (albeit, hopefully provocative) whimsy, it is supported by the logic of time.

In his book, *The Dragons of Eden*, astronomer Carl Sagan invents a way of describing the passage of time since the origins of the known universe. As a point of beginning, Dr. Sagan accepts the "Big Bang" theory: the notion that some 15 billion years ago an explosion involved all the matter and energy in the present universe. Our galaxy— the Milky Way Galaxy—evolved from that explosion, as did the other 100 million galaxies within the limits of electronic and telescopic observation.

Sagan compresses that 15-billion-year time span into one of our calendar years, starting with January, ending 12 months later with the last 24 hours of the last day of December. On such a scale, every billion years of Earth history corresponds to about 24 days of the Sagan "cosmic" year. One second of the cosmic year corresponds to 475 years. On such a scale, during the time it takes you to say "one-a-thousand," nearly five centuries would pass.

Were you to look at the cosmic calendar and begin a search for the date of the arrival of the first humans on the planet, you would begin that search on the last day of December. It was not until the first day of that month, 11 months after the cosmic year began with the Big Bang, that significant amounts of oxygen began to develop in the earth's atmosphere; and not until late in the evening of the last day, about 10:30 P.M. on December 31, that the first humans stood erect on two feet.

In an entire year of cosmic history, Man occupies only the last hour, and it was only during the last minute of that last hour that Man became intelligent enough to plant crops. Lucretius lived during the last four seconds, and the discoveries and technological developments since

HOW MUCH ENERGY DOES IT PRODUCE?

The sun produces more energy in one second than we have produced in all of human earth-history. In three days, the sun sends out more energy than we could create by burning all the wood, gas, oil, and coal stored in our earth.

HOW HOT IS IT?

At its core, the sun's temperature reaches 15 million degrees Celsius—a number almost too huge for our minds to grasp. The English astronomer James Jeans once figured that if we took a pinhead-size piece of matter from the sun's core and put it on earth, its heat would kill a person 94 miles away.

the Renaissance have occupied a fraction of the last second of the last hour of the year's last day. All of recorded history fills the last ten seconds of New Year's Eve.

Yet we begin our second cosmic year (on Sagan's calendar) with large portions of the scientific community claiming that such and such is *fact*, that so-and-so has been irrefutably established as such. Perhaps we need this sort of arrogance. By claiming we have discovered a great deal in one second of cosmic time about a universe which has taken a year of that time to evolve, we can eliminate many unknowns. For those who view unknowns as causes for anxiety, for those who see the cosmos as a frightening, lonely vastness in which we are but windblown specks of dust, the reassurances of such knowledge may be essential.

A second or two ago on the cosmic clock, the universe was too awesome for people to contemplate directly. Myths were required if any order was to be made of the apparent chaos of the heavens, if there was to be any understanding of such elemental upheavals as earthquakes, tidal waves, and eclipses. Because these phenomena could not be explained literally—today, we would say "scientifically"—they were explained fancifully. Some of those flights of fancy taken by the mythmakers have stayed aloft through the centuries; we still speak of Apollo and his fiery chariot, of Jason and the golden fleece.

But we only relate these myths, we do not believe them. Today, we find our sustenance in science. We have the myths of technology. No longer is the sun drawn across the heavens by Apollo's horses; it is pushed and pulled by the forces of gravity. We believe in physics, not fables. Yet because our physics are also our fables, and because the vastness of space and the universe are still quite overwhelming, we are reluctant to debate our "scientific truths." Without them, and without myths, we will be left totally vulnerable to mystery. Thus we hold to our technology as absolute.

Yet it is also logical to assume that during the second cosmic

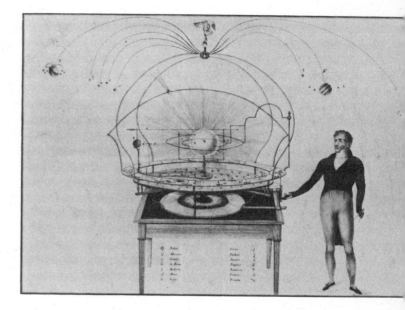

A nineteenth-century French model of the solar system.

In this sixteenth-century German woodcut, a seeker after knowledge pokes his head out beyond the visible universe to seek the knowledge of the hidden realms which lie beyond the workings of our world.

HOW DOES THE SUN WORK?

The exact nuclear process by which the sun brings to birth its enormous energy is not known, but scientists think that it is by a fusion of hydrogen into helium. According to one theory, about every 7 billion years each hydrogen nucleus in the sun's core crashes into another hydrogen nucleus, producing helium. A small fraction of the total weight of the atoms is left over from the nuclear fusion, changing into nutrinos and pure energy, which leave the sun and shoot into space.

The energy released in the sun's core by its thermonuclear reactions takes up to 50 million years to jostle its way to the surface. Near the solar surface, the radiating energy becomes turbulent. At the level of the photosphere, it finally becomes visible as the white ball of light we see moving across our sky.

The sun beams out its energy in electromagnetic waves which bring heat and light to earth. Light leaving the surface of the sun strikes plants on earth a mere 8⅓ minutes later, having raced through 93 million miles of space. That distance is nearly 400 times the space between earth and the moon. At 1,000 kilometers per hour, it would take humans 18 years to fly to the sun in a jet liner.

Because the earth is such a small target for the sun's rays, astronomers calculate that we receive only 2 billionths of the total energy sent out by the sun in any minute. The rest of the energy streams out into space, some of it to be absorbed by other planets, the rest to purposes we know little about.

year to come, there will be evolutions, discoveries, and metamorphoses as elemental as those during the one just past. We are living through a mere moment of discovery—a tick of the cosmic clock that compresses time so that the solar-carbon investigations of a Roman architect and a twentieth-century physicist occur simultaneously.

Recognizing this aspect of the development of humanity, the case can be argued that in a rush to alleviate cosmic anxieties, we have attached too much significance to available scientific discovery. We are telling ourselves that everything that has been learned during the final second of our first cosmic year is, indeed, fact. Since the start of the Industrial Age, the advent of technology a bit more than a century ago (on conventional calendars), we have tended to dismiss myth, to ridicule superstition, to restrain fancy. Thus we argue that the universe is this, the cosmos is that, the solar system is such and such and the sun so much hydrogen and helium at such and such a temperature and mass.

In the so-called developed nations of the planet—and the United States is the foremost of these—the advance toward a technological society has caused a concurrent withdrawal from what can be called a "natural" society. Those who can explain the wind as thermal currents, those who have been taught that the sun is a finely controlled nuclear furnace tend to want to cling to such bits of scientific truth, rather than to ponder the existence of equally apparent natural mysteries.

Whatever eliminates or minimizes mystery also eliminates and minimizes anxiety. In the last half of the last second of the last day on Sagan's cosmic calendar, we have tried to weave a protective shield from the first, fragile strands of scientific discovery to be woven into the tapestry of time. For creatures who lived the 5,400 previous seconds in competition with and in fear of the natural presences we now shield ourselves against, the posture is understandable.

It is also costly. At the same time that shields protect, they also limit vision. They are cumbersome to carry, they are restrictive; the price

HOW DENSE IS THE SUN?

On an average, the density of the sun is about a quarter that of the earth. Its outer layer, the photosphere, is so diffuse that we would call it a vacuum here on earth. Only halfway into its center does the sun become as dense as water. Despite the thinness of the gaseous outer layers of the sun, we can see no deeper than this opaque area. From earth, these gases give a clear-enough outline to make the sun look solid.

The most beautiful and most profound emotion we can experience is the sensation of the mystical. It is the sower of all true science. He to whom this emotion is a stranger, who can no longer wonder and stand rapt in awe, is as good as dead. . . . The cosmic religious experience is the strongest and oldest mainspring of scientific research.
Albert Einstein

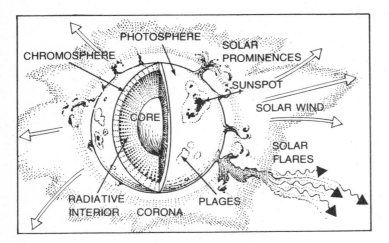

Chromosphere–visible during eclipses or with special telescopes.

Core–center of the sun where hydrogen is converted to helium under great temperatures and pressure.

Corona–visible during eclipses or with special telescopes; ionized gases.

Photosphere–very bright, visible disc on which sunspots and plages occur; marked by strong, turbulent motions, like boiling liquid. Temperature about 11,000°F.

Plages–bright, hot areas on photosphere.

Radiative interior–area through which the sun's energy begins to move outward by radiation.

Solar flares–sudden release of tremendous energy from near a sunspot; spews forth charged particles, rays, and ionized gases that affect the earth environment within a few minutes and for days afterward; can disrupt long-distance radio and telephone communications.

Solar prominences–condensed streams of cool gas within hot corona.

Solar wind–streams of charged particles.

Sunspot–cool, dark areas centered in strong, local magnetic fields; last a few days to 3 months; umbra is dark core; penumbra is outer area.

of their security is the maintenance of the status quo. It can be argued— and it is one of the purposes of this volume to do so—that the shield of technological knowledge so recently acquired and so staunchly maintained by the majority is preventing us from seeing the sun in all its dimensions. From that premise, it can be further argued that such a limited vision of the sun prevents us from living with it in the fullest harmony possible. True, we know more today, in scientific terms, about the sun, than the citizens of ancient Rome for whom Lucretius wrote his poetry. But, in many other, equally important ways, we know less.

As we embark on the first half-second of our new cosmic year it would be nothing less than prudent to correct the imbalance that has occurred during the final half-second of the one just past. There is no reason to be afraid. Why should metaphysics be more frightening than physics? In addition to comprehending the sun's anatomy, we must also be willing to explore its myths. Along with analyzing the effect of ultraviolet rays on the cells of our skins, we should also examine more closely the effects of sunspots on our psyches. And, rather than using the wealth of scientific detail accrued about the sun as a block to our further contemplation of the sun's blessed mysteries, we should acknowledge the fragility of those "truths" and ponder the immortality of mystery.

It is time we did so. Time, because one of the sun's legacies— the millions of tons of fossil fuels, stored months ago, not seconds ago, on the cosmic calendar—has been explored enough so that its limits are now within definition. This comprehension of limits is a shattering discovery—as disturbing in the twentieth century as the recognition in the fifteenth century that the earth was not the center of the universe. The Industrial Age began with the certainty that natural resources were there for the taking. In a few generations—the flicker of an eyelash by the cosmic clock—we have come to understand the finite nature of iron ore, oil, natural gas, bauxite, coal, and the other raw materials with which a century of industrial expansion and exploitation has been energized.

 *SUN FACT*

SOLAR STATISTICS

AGE

At least 4.5 billion years

TEMPERATURE AT CORE
15 million degrees Celsius

TOTAL ENERGY OUTPUT
3.83×10^{23} *kW*

DENSITY AT CENTER
90 × density of water

AVERAGE DENSITY
1.41 × density of water

DIAMETER
864,000 miles (109 × earth's diameter)

VOLUME
337,000 trillion cubic miles
(1,300,000 × earth's volume)

SURFACE GRAVITY
28 × earth's gravity

AVERAGE DISTANCE FROM EARTH
92,900,000 miles

PERIOD OF ROTATION
26.8 days (equator), 31.8 days (poles)

SPEED OF SUNLIGHT
93 million miles to earth in 8¹/₃ minutes

CHEMICAL COMPOSITION
OF PHOTOSPHERE
Hydrogen   73.5%
Helium   24.8%
Oxygen   0.8%
Carbon   0.3%
Iron   0.2%
Neon   0.1%
Nitrogen   0.09%
Silicon   0.07%
Magnesium   0.05%
Sulfur   0.04%
Other   0.1%

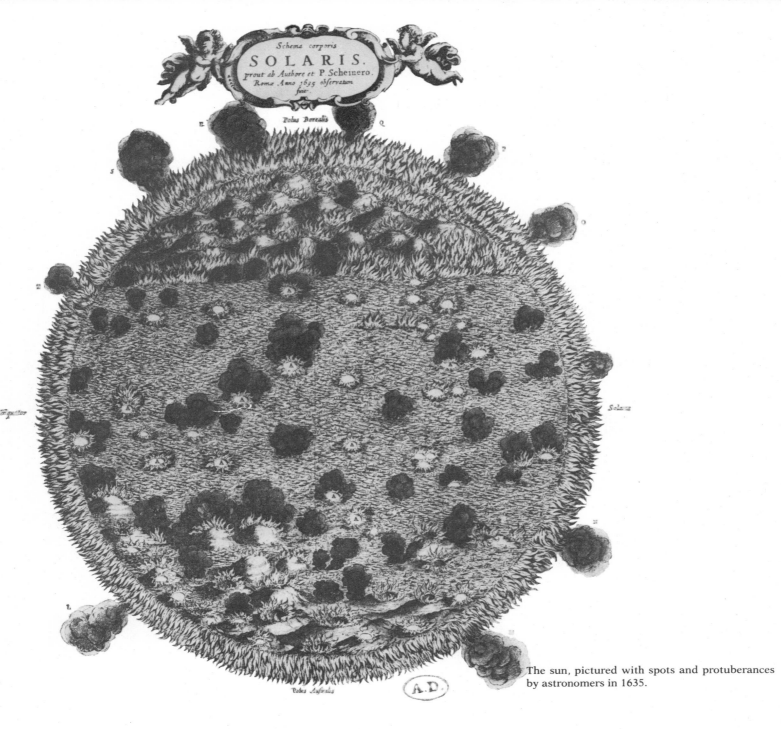

The sun, pictured with spots and protuberances by astronomers in 1635.

Immissione Refractoria composita.

Seventeenth-century observations of sunspots were made as they are today. Here, Christopher Scheiner and a fellow Jesuit trace sunspots projected by the telescope onto a card. By this method they could note the changes and motions in the spots.

Thus the sun not only warms and illuminates the planets, it also envelops them in its own atmosphere. We might be tempted to picture the sun as a mother hen spreading her wings protectively over her chicks. But the moment such a thought entered our heads, we would probably dismiss it again; after all the image seems a little far-fetched and anthropomorphic. Yet in fact everything astronomers have learned in the last decade indicates that this far-fetched analogy is really quite appropriate! Without the protection afforded us by the solar atmosphere, our earth would be uninhabitable.

Hoimar von Ditfurth, *Children of the Universe*

The remaining dimensions of those resources are, philosophically speaking, relatively unimportant. To an industrialist whose empire stands on the width of a copper seam in Zaire, specifics count. To a human being contemplating a future of limits, such specifics are irrelevant. What does it matter if, at current rates of consumption, telephone companies have only enough copper to last 22 years, or possibly 25? What matters is the concept of limits defined by such realities.

To a person conditioned by the limitless notions of the Industrial Age, the specter of limits can become a frightening image. As with

[*Continued on page 27*]

Sunspots—cool, dark areas centered in strong magnetic fields—have been observed for more than 2,000 years with the naked eye. With the more accurate telescopic observation available since Galileo, scientists charted their movements to show that the sun rotates in a period of about 27 days. The spots develop in 11-year cycles and their changing magnetic fields have been connected with cyclic changes on earth.

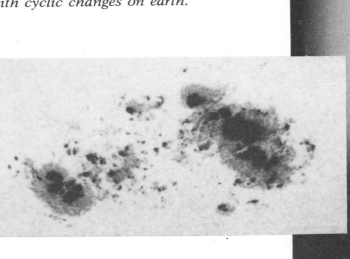

The sunspot of April 7, 1947

24

Enormous solar eruption of helium shoots 500,000 miles into space.

The surface of the sun is far from quiescent. Vast prominences leap from it, sometimes as far as 300,000 miles into space. Flares brighten small areas and then subside. Dark sunspots, looking much like whirling solar tornados, form and grow, eventually breaking apart. Hovering above the surface is the ghostly solar corona, with a temperature of more than one million degrees. From it the sun throws out a varying flux of subatomic particles and magnetic fields that expand through the entire solar system as the tenuous solar wind.

Trudy Bell

Clouds of helium gas blast out 350,000 miles from the sun in a solar energy spectacular.

most ghosts, the fright originates in the eye of the beholder, not within the apparition. There is no absolute cause for anxiety in the reality of limits. There are ways to live with the reality; indeed, some would claim that such lives might well be more fulfilling than many of those lived during days of mass production, mass consumption, and mass waste. Seen through the eyes of a person who has made peace with limits, those same hallmarks of the Industrial Age might be converted to community production, careful consumption, and almost no waste. Such a design is, after all, much more in keeping with natural rhythms. Indeed, in the world of living cells, only one, the cancer cell, follows a pattern of exponential growth.

But such arguments do little to reassure the citizens of the Industrial Age. Presented for the first time with irrefutable proof that limits are an unavoidable aspect of the grand design, many of them, most of them it seems, grow anxious, frightened, and begin a desperate search for ways to continue the ever-ascending line of exponential growth. They press for quick-fix solutions like nuclear fission and fusion.

It is to be expected that technology's children would look to technology for protection from the imagined threat which a concept of limits poses. Yet it is also ironic that this devotion to technology could blind them to alternatives that may prove more practical, more benign, and more compatible with the planet's inherent natural balance.

In an age when limits must be acknowledged, there is no presence more reassuring than the sun. The problem, however, is that the time of technology which gave us the Industrial Age also nurtured science as an almighty. Thus, for the past several decades, our image of the sun has been created by the innards of radio telescopes, the lenses of spectroscopes, and instruments for recording the density of cosmic rays. For at least two generations, the sun has so increasingly become the property of science that it has all but vanished from popular view. This has led to a double-edged phenomenon: on the one hand, a large group of scientifi-

"Discovery with parhelia (solar halo) behind"; by Dr. E. A. Wilson, who died with Scott in the Antarctic.

27

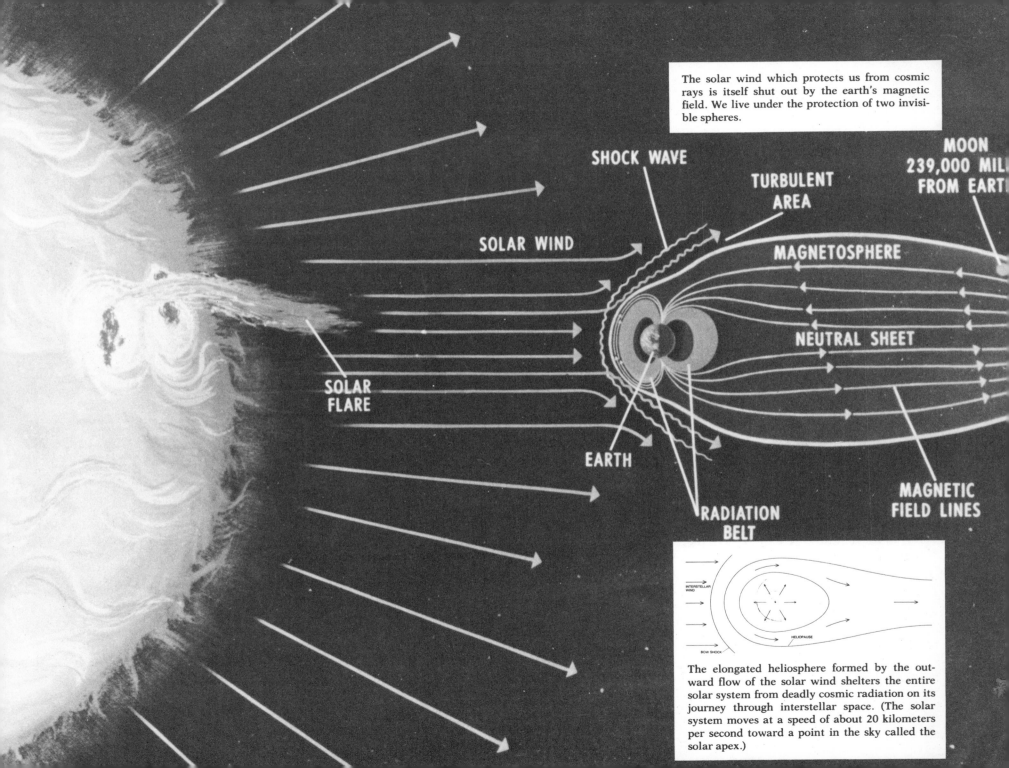

The solar wind which protects us from cosmic rays is itself shut out by the earth's magnetic field. We live under the protection of two invisible spheres.

SHOCK WAVE

TURBULENT AREA

MOON 239,000 MIL FROM EART

SOLAR WIND

MAGNETOSPHERE

NEUTRAL SHEET

SOLAR FLARE

EARTH

RADIATION BELT

MAGNETIC FIELD LINES

INTERSTELLAR WIND

HELIOPAUSE

BOW SHOCK

The elongated heliosphere formed by the outward flow of the solar wind shelters the entire solar system from deadly cosmic radiation on its journey through interstellar space. (The solar system moves at a speed of about 20 kilometers per second toward a point in the sky called the solar apex.)

cally oriented astronomers believe they have recorded most of the basic sun science that there is to be learned; on the other, masses of nonscientists have grown further and further away from the sun as they spend ever-increasing amounts of time in the artificial environments which urban and suburban technology have created. For a species which has lived all of its brief cosmic life in close contact with natural presences, this sudden breakup of its relationship with the sun can prove traumatic; in the eyes of some observers, it already has.

It should be, according to those who argue the case, beneficial to us all, scientist and layman alike, to re-establish the sun's complete identity. We must learn—as our recent ancestors surely learned—to see the sun in each of its many dimensions. Once we come to comprehend the wondrous complexity, the potent mystery, of our star, once we become re-acquainted with its traditions as well as its dimensions, we will not only begin to "see" the sun more clearly, but we are quite likely to learn that living in harmony with the limits of this planet is no cause for anxiety. Instead, it can be cause for contentment.

According to its proponents, a solar village's dwellings could be energized by the sun, as well as some small manufacturing plants and several offices. Residents would travel in electric vehicles charged by solar devices, and the village would include a solar thermal-electric plant, systems for recycling sewage into fertilizer, greenhouses, and a 100-acre, organic vegetable farm.

Whatever else a solar village could prove, or disprove—and the arguments are continuing—it could give all of us an indication of how we will, at least, re-establish our acquaintance with our star. It may not happen in your state, but somewhere in this nation before the century has ended, a solar village will bloom. Its residents will live in harmony with the sun, their energy will come from the sun, the patterns of their lives will shift from designs of incremental growth to curving lines of equilibrium. The schools their children attend will include a solar curricula, and each and every citizen will sense a profound change in the rhythms of

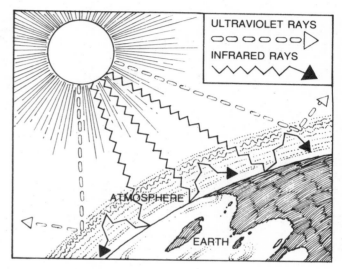

Greenhouse effect.

The earth's atmosphere is a delicate part of our relationship to the sun, serving us as a shield and also a blanket. It protects us from destructive ultraviolet and x-rays while letting through and then trapping the solar heat of infrared radiations, without which we would freeze.

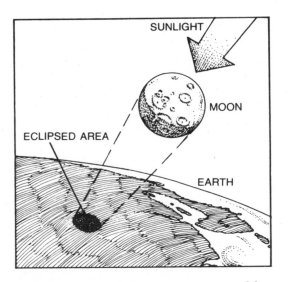

Eclipse of the sun by the moon interposed between sun and earth.

The brilliant corona of the sun seen at the moment of total eclipse is one of the most striking sights in nature. Pictures of the corona (Looking directly at the sun during an eclipse damages the eye, so the corona is observed indirectly.) suggest that strong magnetic fields existed on the sun before we could actually measure them. Now we know that the magnetic fields at the sun's poles equal those of earth's; the fields of sunspots and other active areas are thousands of times stronger.

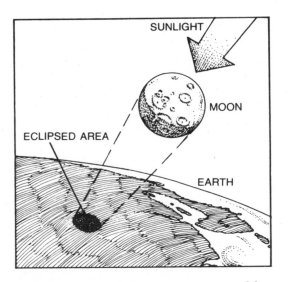

their lives as they begin to order those lives in conjunction with the sun.

As the prototype community begins to evolve, it will be watched by hundreds of millions of other Americans still living in Industrial Age villages, towns, and cities. And as those millions come to comprehend the benefits their solar brethren are discovering, the move toward national solarization will become a social revolution as fundamental as any in the history of civilization.

If you tend to dismiss such claims as visionary rant, then you should note two items. One, the reliance on prototypes that will usher in the solar villages is a reliance that has grown out of the scientific approach, the hallmark of the Industrial Age. Thus, we will take the best of the past and apply it to developing a more fulfilling future. We have learned how to measure and scale prototypes, and one of the fine advantages of solar models is the very diffusion of the sun's energies. We know they are bestowed over the entire planet, and even more generously, over what we perceive as space. We are able to calculate with considerable accuracy just how a solar village that works well in Los Angeles will also work in Salt Lake City, Sioux Falls, and Bangor. We have become skilled collectors, communicators, and transposers of data; that's one reason a solar village will be emulated in geometric progressions.

The other is its basis in science. The solar village is, in fact, the only justification I can see for the incredible accumulation of solar science that is now at the disposal of the earth's technicians. That accumulation of data, that assemblage of theory and hypothesis, that great mass of solar information that has begun to be gathered, each of the collected bits of sun history that have been discovered and recorded indicate that we can live by the sun. There is, according to scientific sun documents, ample solar energy to sustain us all. What is needed to allow that science to be usefully applied, of course, is more public awareness, more public trust in the sun's dimensions as a provider.

Which is not to say that science should be either resisted or

 *SUN FACT*

One of the most traumatic changes that life can experience is the sudden and unexpected disappearance of the sun. On the rare occasions of total eclipse, living things are thrown into complete confusion. I have seen an eagle drop straight out of the sky to take refuge in the crown of a tree, and a foraging troop of baboons rush into the defensive formation they usually assume in response to a predator, neither species knowing quite which way to turn to meet this new and unexpected threat. Only man knows when to expect the next eclipse of the sun by our moon, but all life is tuned to the daily obliteration of sunlight by the movement of our own planet.

Lyall Watson, *Supernature*

Photograph of total eclipse of June 8, 1918, Green River, Wyoming.

31

Scientists have discovered what the sun is made of by deciphering the mystery of dark lines seen running across the bands of rainbow color when the solar spectrum is looked at through an instrument called the spectroscope. William Wolleston, an English physicist, first noted those bands in 1802. A decade later Joseph von Fraunhofer of Munich had recorded more than 500 of them in the spectrum, and today we know they are almost countless. Scientists think the dark lines are an effect of the absorption of some light rays from the solar interior by like elements in the sun's outer atmosphere. The black lines let a scientist read off the elements as exactly as if they were written in formulae.

minimized. Lucretius said, ". . . things can not be reduced to nothing. . . ." It is an argument to keep in mind if you are inclined to dismiss too much of science in an attempt to balance the excesses of the Industrial Age. Science should not be reduced to nothing; instead, the best of its insights and the best of the technology it has created will be essential if we are to re-establish a mutually beneficial relationship with the natural presences of the universe.

When that renaissance begins—as it is most likely to—with the evolution of a new and rewarding relationship with the sun, then solar science will be one of the elements of that relationship. However, like sun myth, sun superstition, sun magic, and sun lore, sun science is more than a mere collection of data gathered over the years like so many ears of corn from a field, to be shucked and stored in a crib.

The sun is too mysterious for that. You can, if you wish, set out to disprove many of the "scientific facts" known about the sun: it is reasonably probable that you will be successful—not popular, but successful. First, you must read most of the innumerable descriptions of the sun written by scientists and men and women concerned with the "nature of things" since Lucretius and before.

Each of them illuminates some aspect of the sun, but all of them underscore its essential mystery. And we end up as far from certainty, although we "feel" closer to it, as some of the astronomers writing centuries ago. Consider this report published in the *Boston Globe* late in 1979:

A long, careful search through centuries-old records in England is yielding the surprising conclusion that the sun is shrinking.

Solar astronomer Jack Eddy says the sun seems to be losing about 5 feet of diameter every hour.

Although many of his fellow astronomers are

Skylab, shown here with the earth's atmosphere as background, let us for the first time observe the sun free from the distorting effects of that atmosphere.

Prism spectroscope used with telescope ca. 1880.

A solid tank of dry-cleaning fluid, a trap for solar nutrinos at Brookhaven National Laboratory, is giving signals of a new solar mystery. The nutrinos (subatomic solar particles that travel at the speed of light from deep within the sun's core) could give us information about the sun only 8⅓ minutes old. But the particles measured by the tank don't square with the number projected by our solar theory: sometimes they're many fewer—¹/₃₀th the number projected—; at other times they match up. Could the sun be turning off and on? We don't know yet.

finding this idea hard to swallow, Eddy insists that's what the records seem to show.

"This really shouldn't surprise us," Eddy said in an interview. "After all, every star we can see was formed by shrinking. The sun itself was formed in a process that involved shrinking."

Neither he nor other astronomers can say what the effect of a shrinking sun might be on the earth's climate. So far there seems to be no evidence that the supposed decrease in the sun's size has made any discernible difference to people on earth.

Nonetheless, other scientists are dismayed at the very idea of a diminishing sun, since much of modern astronomy is based on observations of the sun, which has long been used as astronomy's steady, reliable "standard star." Eddy is saying, basically, that the standard maybe isn't standard—that their favorite yardstick is shrinkable.

Eddy, from Boulder, Colorado, delivered a paper on his conclusions yesterday at a meeting of the American Astronomical Society at Wellesley College. He is visiting this year at the Harvard/Smithsonian Center for Astrophysics in Cambridge.

After several years of work, both at the famed Greenwich Royal Observatory in England and at the US Naval Observatory in Washington, Eddy said he

Impression soleil levant by Claude Monet.

As a child I used to visit a neighbor's house where glass prisms, probably saved from some old chandelier, were left lying on the window bars. On fine days they threw a dozen or more little colored patches, running from blue through green and yellow to red, on walls and ceiling, on furniture and sometimes on one's own clothes and skin. As we played, or worked at our sewing class, all these delicious patches wheeled steadily around the room. One which was full on a clock face when it said three o'clock would be sliding off its edge by a quarter past.

It still seems to me extraordinary that everyday light is built up from this dazzling range of colors; still more extraordinary that these colors are due to the burning of the chemical elements composing the sun. Each element burns with its own particular wave lengths and therefore with its own particular colors. Salt or any compound of sodium, for example, burns with two wave lengths that make a bright yellow show. So all the

[Continued on page 37]

has to conclude that the sun's diameter is shrinking, and has perhaps been shrinking for 400 years or more. He reached this conclusion, he said, after first carefully searching through the Greenwich records, the oldest, longest continuous set of solar observations in the world.

These daily observations at Greenwich, which began in 1750, Eddy said, show that the sun's diameter has shrunk—and is shrinking—at the rate of two arc seconds per century. That amounts to 830 miles per century, or 8.3 miles per year, or 120 feet per day.

"I began looking at it (the Greenwich records) because I was looking for something that we haven't found yet; that the sun's diameter might change as it goes through long periods of change in solar activity," Eddy explained.

The original Greenwich observations were taken as astronomers tried to tie down the Earth's slightly variable rate of rotation with some precision. The goal was to determine the position of the sun's center exactly. In doing this, however, they provided Eddy and others with a long-term record of the sun's diameter.

What these records showed, once they were dumped into a computer for analysis, was a steady, gradual decrease in the sun's diameter.

Eddy's fellow astronomers, however, haven't embraced this idea with huge enthusiasm. Dr. Paul C. Joss, at MIT, commented:

"I think if Jack is right, it's really an earth-

colors of the rainbow, and indirectly, all the colors of art and nature, are the product of the sun's chemistry.
Jacquetta Hawkes, *Man and the Sun*

Voyage of Life: Old Age by Thomas Cole.
Sunlight breaking through the heavily banked clouds symbolizes a heavenly destination in this nineteenth-century painting. Sunlight as the image of the source of life, comforting presence, understanding, goodness, and power permeates our symbolism long after we have ceased to tell stories about our relationship to the sun.

shaking development. It means we have no current understanding of what the sun's doing."

Eddy noted, however, that he suspects the shrinking is only going on in the sun's outer layers and that the sun's interior regions are probably not collapsing.

"If it's just a surface effect," Joss said, "then there may be a way out," but it would still be difficult to account for the amount of heat coming from the sun through such a process. "Basically," Joss said, "I'm very skeptical at this point."

Dr. Irwin Shapiro, also at MIT, commented that he finds Eddy's suggestion "extremely unlikely."

Eddy is proposing, too, that discovery of the sun's shrinkage might provide an answer to one of the most interesting new problems in solar physics.

Scientists call this the problem of the missing neutrinos. Careful measurements show that only one-tenth as many tiny particles called neutrinos are coming from the sun's interior as expected.

If these measurements are correct, estimates of the sun's interior heat may have to be revised downward from 15 million degrees to 14 million degrees. This would mean that some nuclear theory—about how nuclear processes run in the sun—might have to be rethought.

If, however, the sun gets some of its energy from shrinkage, plus some of it from nuclear fires, then the problem of the missing neutrinos may be solvable.

To my eyes, one of the sentences that leaps from this account comes from Dr. Paul C. Joss of the Massachusetts Institute of Technology: "It means," he says of astronomer Jack Eddy's work, "that we have no understanding of what the sun is doing." That observation, by a scientist speaking in the halls of one of the nation's foremost scientific institutions, takes us back to Rome more than 400 years ago when the Cardinals of the Church used almost the identical words in response to Galileo's observations that the sun, and not the earth, occupied the center of our solar system.

Science is like that. Generally considered to be the most absolute of pursuits, science, as its history proves, is one of the most volatile. Its basic structures change with each new discovery; its theories are rewritten as much by philosophical insights as by the revelations of a microscope. How much, after all, could we expect scientists to be able to

Morning Sun by Arthur Dove.

The universe, then, is not that empty, cold, and lifeless space whose majestic indifference once inspired us with fear as well as admiration. It is the living soil from which our earth has sprung; and our planet is joined to it by a thousand roots. We are not some tiny oasis permitted to exist in a hostile universe solely because of our insignificance. Instead, our earth can be shown to be a focal point where various cosmic powers conjoin to fashion a living world.

Hoimar von Ditfurth, *Children of the Universe*

The Vedas say, "All intelligences awake with the morning. . . . To him whose elastic and vigorous thought keeps pace with the sun, the day is a perpetual morning. It matters not what the clocks say or the attitudes and labors of men.
Henry David Thoreau

tell us, when, like all of us, they are working and living in a time frame as fleeting as the blink of a cosmic eye?

They are not charged with giving us the history of the sun; they are charged with observing our star during their lifetimes, and recording those observations with scientific precision. We should not be disturbed by their "new" discoveries. Instead, we should welcome them as an indication of how many of the sun's wonders remain to be explored.

What we, and the would-be builders of the nation's first solar community, have to build on is the total body of solar science that has been accumulated over the centuries. What that tells us is that the sun, like every other natural presence, is alive, is evolving, is vast, is still mysterious, and is surely capable of providing this planet and others with the energy to sustain our lives, and the lives of countless generations to come. What applies to other natural presences applies to the sun: it must not be abused, taken for granted, or carelessly exploited. That is what is known about the sun, scientifically. The same principles apply to the human mind, another natural presence which generations of scientists have claimed to comprehend.

We have learned there is a great deal left to learn about the human mind; we are discovering that there is a great deal left to be discovered about the sun. Yet the existing body of our scientific solar discoveries has established enough of the sun's energy dimensions to allow us to think realistically about living in a solar society. That is, if you will, a scientific fact. A solar community could, indeed, function, and function well.

It could have already been built had we achieved a fuller understanding of our star. If we had coupled some sun fancy with the sun facts our scientists have recorded, if some of the bright threads of legend had been woven into our contemporary solar tapestry, if we had re-acquainted ourselves—as children of the Industrial Age—with the solar myths that were part of the fabric of life before the steam engine, before

the computer and the incandescent bulb, then we would, even now, be building our solar communities.

Science does not build communities; imagination does—imagination energized by scientific observation. That observation of the sun, as we have seen, has been documented for centuries. And although new scientific discoveries are still to be made on the sun's vast and fiery plain, enough have already been reported to give us all the sun facts we need to start building a solar society. What we lack, after two centuries of taking energy from beneath the earth, is faith in the sun—a faith that science alone can never create. To do that, fact and fancy must work together.

"AS I SEE IT, IT'S TECHNICALLY FEASIBLE, BUT COSTLY AND NOT READY FOR COMMERCIAL APPLICATION"

Chapter 2
SUN FANCY

Praise to thee, my Lord, for all thy creatures,
Above all, Brother Sun
Who brings us the day and lends us his light.
SAINT FRANCIS OF ASSISI

2   *SUN FANCY*

If the challenge were to grant a comprehension of the sun to one who had lived underground since birth, and there was but a brief, limited time span allowed, then I would opt for a reading of sun myths, rather than an attempt at a scientific explanation of the solar presence. All of us might find much to be learned from such an approach to the sun. Like philosophers of centuries past, we might re-establish (and be the better for it) the links between the solar system of our external universe, and the mirror of that system that is our internal mystery. Our consciousness, our being, our presence, our soul, our spirit, whatever name is given to the essence of our being, is, like the sun and the universe, mostly mystery—a bright star, able to be "seen" by each of us, yet fully understood by none.

The sun within, as well as the sun without—people of every continent, of every nation, have been aware of the sun within themselves, have been aware of the awesome differences of light and dark, good and evil, intellect and brute, wisdom and ignorance, miracle and black magic, heaven and hell, life and death. From these contrasts, the yin and yang concept of their world, they have spun sun myths and sun legends ever since the first heroes and villains were described by the storytellers and temple priests of the earliest known settlements of the Sumerians and Mesopotamians—25 seconds from midnight on the last day of December on Sagan's cosmic calendar, some 2,000 years before the birth of Christ on the earth's calendar.

Some 3,500 years later as Galileo and the Cardinals clashed,

Myth is the ever changing mask that the mind of the beholder fits over a reality he has never truly seen.
Stephen Larsen, *The Shaman's Doorway*

45

the waning of mythical cosmology began. Bertolt Brecht, the German playwright, caught the spirit of the change in these lines from his play, *Galileo*. The old Cardinal, the astronomer's opponent, resists the advance of science with these words: "So, you have degraded the earth [by stating that the sun is the center of the universe] despite the fact that you live by her and receive everything from her. I won't have it. I won't have it! I won't be a nobody on an inconsequential star briefly twirling hither and thither. I tread the earth. The earth is firm beneath my feet, and there is no motion to the earth and the earth is the center of all things, and I am the center of the earth, and the eye of the creator is upon me. About me revolve, affixed to their crystal shells, the lesser lights of the stars and the great light of the sun, created to give great light upon me that God might see me—Man, God's greatest effort, the center of creation. . . ."

In the four centuries since, the old Cardinal's battle has been all but totally lost: science has all but obliterated myth. And yet, during the same decade that the first spacecrafts ventured on their tenuous voyages beyond the earth's atmosphere in what is surely the most universally acknowledged scientific triumph, myths resurfaced among the populace. *The Lord of the Rings* became a best-seller; *Jonathan Livingston Seagull* mirrored the legends of the earliest Americans: the Cree, the Blackfeet, and the Sioux. Even as solar energy powered the transmitters that allowed the astronauts to describe outer space to billions of earth-bound spectators, mythical devices were raised from the past and used, as they have always been, to help a bemused humanity try to explain celestial and global occurrences for which "logical" explanations were proving increasingly inadequate. In spite of its seven-league advances, science in the twentieth century's second half seems to be raising more questions than it answers, and scientists—in spite of their allegiance to empirical fact—seem more and more to disagree, to state, on the one hand, like Jack Eddy, that the sun is shrinking, and, on the other, like Dr. Irwin Shapiro, that Eddy's theories are "extremely unlikely."

Alchemy, the search for the unity of physical elements, was also a mystical search for the unifying principle of all things. Here an alchemist stands with the sun shining in his window in unity with the moon. Near him is the astrologer's sun sign, Leo the lion.

The ancient astrological symbol for the sun, ⊙, portrayed the place of the individual in the circle of the whole, of spiritual reality. As an image of the center around which other matter in the system revolves, it also symbolizes the organizing principle for single atoms as well as galaxies.

An astrological chart is a symbolic map of the solar system against the backdrop of the 12 constellations as it appeared from earth at the point of our birth. As the sun is the central life source in the universe, so astrologers say that the "sun sign," or place of the sun at the moment of birth, shows the basic individual lifeforce.

In his book, **Astrology for the New Age,** *Marcus Allen traces the rhythm of the sun's journey through all the signs of the zodiac. In Aries (March 21 through April 20), we feel the rush of spring life, while in Taurus (April 21 through May 20), we feel the deeper, slower rhythm of roots growing grounded. For each season, we feel the intense energy of beginning, an unfolding, settling time, and then another high-energy time of change.*

Germanic Sun Idol From the *Sachsisch Chronicon*, 1596.

The circles of stone at Stonehenge provided sighting of the sun at summer and winter solstice and may have been a ceremonial place for worship of the rising sun. The outer ring of stones may have been an eclipse computer.

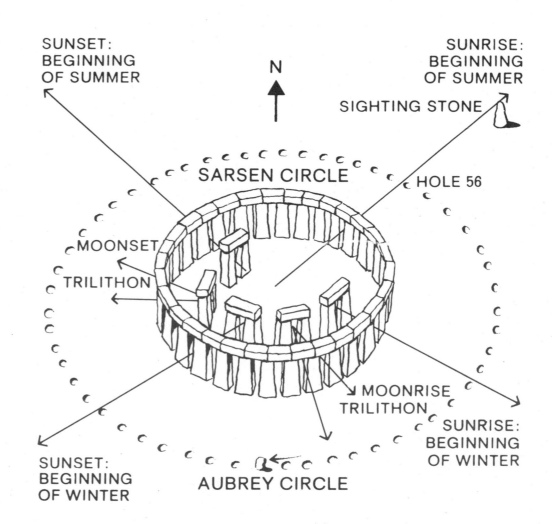

SUNSET:
BEGINNING
OF SUMMER

N

SUNRISE:
BEGINNING
OF SUMMER

SIGHTING STONE

SARSEN CIRCLE

HOLE 56

MOONSET

TRILITHON

MOONRISE
TRILITHON

SUNRISE:
BEGINNING
OF WINTER

SUNSET:
BEGINNING
OF WINTER

AUBREY CIRCLE

The confusions are honest. Ever since the first sunrise, we have sought what we term "logical" explanations for natural phenomena. But logic, pinned as it often is, to scientific facts that shift with the tides of discovery, is often not as popular as myth. And, at the beginning of civilization, before the microscope and before the pocket calculator, there were natural presences so overwhelming in the magnificence of their mystery that they could only be absorbed by translation to a kind of human scale.

It is that translation which created myth. It is the passage of the sun across the sky which became Apollo's fiery chariot, the blazing of the noonday sun over India which, in turn, created a sun-god and a sun-religion. These myths were believed by people because to think otherwise would have meant the contemplation of the unthinkable.

We still have our myths, and for the same reasons. When the Industrial Age was born, it was swaddled in myths of utopia, of a society in which poverty would be erased, diseases cured, and the wrath of Nature turned aside by new, wonderful inventions. We have lived long enough to be able to look back at the arrival of the Industrial Age and ask ourselves: "How could people ever have believed that?" Yet we know that they did, just as we know that the ancients drew their solar myths and their sun-gods from the most wondrous and inexplicable presence in their daily lives—their sun, our star.

The sun is as real for us as it was for the ancients. In ways we can not comprehend—because we now have the myths of science—it was the ruling reality. Thus, with the moon and the stars, with the animals and the trees, with life and with death, with fire and water, the sun became the progenitor of myth, and the parent of religion. In the absence of any believable competition, the sun was quite completely believed. That was the importance of sun myth. If it existed today, the Solar Age would have long since begun.

It seems time, then, to restore mythology to our solar percep-

In the twentieth century, sun stories live on in children's books, a place where wisdom can speak simply. "One day the sun woke up crabby and mean" begins one story by Claudia Fregosi called Sun Grumble. *Looking for something new and exciting to do, the sun races off to play with the stars and kick meteors. The earth, of course, got cold and dark—for plants, animals, and people. Then the sun, seeing what a mess he'd created by his running off, brings back his sunbeams.*

In La Fontaine's fable, the sun and north wind compete in the task of getting a cloak off a young horseman's back. The north wind's roaring blasts failed, for the young man gathered his cloak even more tightly about him. But as the sun let his gentle rays shine down, the horseman grew hot, took off his cloak and dipped into a river for a swim.

The accomplishments of gentle solar heat and the fidelity of an ever-present sun shine through the whimsy, just as the meanings of solar stories have always done.

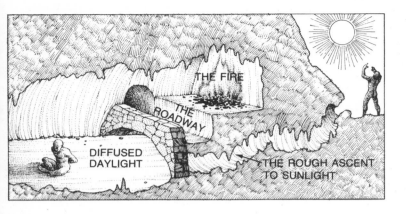

THE FIRE

THE ROADWAY

DIFFUSED DAYLIGHT

THE ROUGH ASCENT TO SUNLIGHT

PLATO'S MYTH OF THE CAVE

In the central book of the **Republic,** *the Greek philosopher Plato compares the ascent of the soul in wisdom to the ascent of a man from a cave into sunlight.*

Suppose, says Plato, that a prisoner chained in a cave was released and led upward over a steep path into the sunlight, would he not be distressed and unable to see even one of the things now called real? He would have to get used to it, first looking at shadows and water and only later seeing things directly by sunlight. Last of all would he look at the sun itself and only then would he understand how it is the sun who provides seasons and is the cause of all that he sees below.

So, says Plato, the rise of our soul toward wisdom is like the gradual movement from the dull light of the cave to the world of sunlight.

tions, this time not as an effort (like the old Cardinal's) to deny science, but to complement it, to use myth knowingly, to add it as yet another dimension of understanding which can bring us closer to the eternally mysterious natural presences which are a part of our existence. Of these, the sun is the foremost presence, and, as we seek to understand and tabulate its energy potentials, we would do well to also re-examine the stories that have been told about it in every corner of the world since storytelling first began.

The sun stories can never be counted. They are as pervasive and varied as the human populations that have peopled the earth for the past 4,000 years. They spring from every culture, every continent, every nation; they are told in every languaee, they live in every history. A book of collected sun myths would be thousands of pages long—all those words, all those images, all those adventures woven around one hero, the invincible sun.

Here are two, chosen arbitrarily from hundreds I have scanned—chosen because they epitomize the style of ancient sun myths, and also because they give the reader an inkling of the diversity—both geographic and metamorphic—that the sun has inspired over the centuries.

The first is from Japan: "After the creation of the world of living things Izanagi [parent of the world] created the greatest of his children in this way. Descending into a clear stream he bathed his left eye, and forth sprang Amaterasu, the great Sun-Goddess. Sparkling with light she rose from the waters as the sun rises in the east, and her brightness was wonderful and shone through heaven and earth. Never was seen such a radiant glory. Izanagi rejoiced greatly and said, 'There is none like this miraculous child.' Taking a necklace of jewels he put it round her neck and said, 'Rule thou over the Plain of High Heaven.' Thus Amaterasu became the source of all life and light, the glory of her shining has warmed and comforted all mankind, and she is worshipped by them unto this day."

50

*High above, he pitched a tent for
 the sun
who comes out of his pavilion
 like a bridegroom
exulting like a hero to run his
 race.*

*He has his rising on the edge of
 heaven,
the end of his course is in its
 furthest edge,
and nothing can escape his
 heat.*

Psalm 19

The Sumerian sun-god Shamash was said to travel across the sky in a chariot whose one wheel was a flaming disk. His piercing eyes could see into the future; his rays could entangle the plans of evildoers. Here Shamash sits with the sun disk on an altar in front of him.

51

Japanese painting, late fourteenth or early fifteenth century. Amida Buddha welcoming souls to paradise. The sun's rays as a symbol of holiness shine out around heads in both Eastern and Western paintings.

The second myth (and both are taken from the book, *Myths of the Sun* by William Tyler Olcott), comes from Mexico, like Japan, a nation which has long held the sun in much esteem. "Nexhequiriac was the creator of the world [so the Mexican myth goes]. He sent down the Sun-God and the Moon-God to illuminate the earth so men could see to perform their daily tasks. The Sun-God pursued his way regularly and unhindered, but the Moon-God, being hungry, and perceiving a rabbit in her path, chased it. This took time, and then she tarried to eat it, but when she had finished her meal, she found her brother, the Sun, had outdistanced her and was far ahead, so that ever thereafter she was unable to overtake him. Since then, the Sun has been ahead of the Moon, and looks fresh and red while the Moon looks sickly and pale. Those who gaze intently at the Moon can still see the rabbit dangling from her mouth."

Aspects of each of the other thousands (surely, tens of thousands) of sun myths are mirrored in these two, brief stories. Like each of the others, they define physical realities, and, at the same time, create metaphysical overtones. They tend, also, to humanize on the one hand, and deify on the other. Indeed, it is difficult for solar historians to separate sun myth from sun legend from sun ceremony from sun religion. The solar presence has touched every culture, and every culture has woven the sun into its heritage, looking upward to the heavens for a being larger than life, deified by its very remoteness, its mystery, and its obvious necessity.

Egyptians in the Sahara and Eskimos in the Arctic—two peoples who would have the most difficulty communicating even now—unknowingly shared a sun myth for centuries. Both have a legend about one of their heroes who rowed the sun at night from the west to the east to make sure it would have the strength to rise again the next morning. But for the Eskimos, it was a female sun who was transported; for the Egyptians, the sun was a god, not a goddess, and the moon was his sister.

It is an indication of the sun's universal significance for the

SUN FANCY

The ancient Aztec temple at Tenochtitlan (now Mexico City) itself was a sun-symbol. Its 360 steps stood for the days of the Aztec calendar, the journey of the sun. The four sloping sides create a feeling of rising up in the sky, closer to the sun and other gods.

In African Dahomey, it is said that the sun and moon are twins. Lisa, the sun-god, is fierce and brave; Mawu, the moon, is gentle and cool so that people love to tell stories under his light. As we need both sun and moon, so the story tells us, we need both strength and wisdom.

Throughout their history, the Egyptians placed the sun-god Ra along with Osiris, god of death and resurrection, at the center of their religious ritual. Akhenaten, becoming pharaoh in 1367 B.C., tried to make the worship of a single sun-god Aton the only religion of his people. His faith in the god survives in the hymn which he composed for his tomb.

[Continued on page 55]

An Egyptian pharaoh with his wife.

ancients that its sex changes with the cultures who have woven the star into their lives and their legends and their worship. In aboriginal Australia, the sun is a woman, the moon a man. Shakespeare speaks of the moon as "she," while in Peru, the moon is a mother and wife to the sun. On the Malay peninsula, both sun and moon are female, and an Indian tribe of South America sees the moon as a man who is married to the sun; the Ottawa Indians of North America describe the sun and moon as brother and sister.

It is said (according to Olcott) that General Benjamin Harrison once called the Shawnee chief Tecumseh for a conference. "Come here, Tecumseh, and sit by your father," said the general. "You my father?" replied the chief, frowning. "No, yonder sun is my father, and the earth is my mother, so I will rest on her bosom." And he sat upon the ground.

Tecumseh, a man of culture, a man of the nineteenth century, spoke of "his father, the sun" as easily as he would speak of his human progenitor; for the sun, to the American Indian, was a presence to be praised. Centuries of custom, legend, ceremony, ritual, dance, and even tribal policies as critical as war or peace were based on the sun.

Like their counterparts in Egypt, Greece, Japan, Central and South America, the early tribes of North America saw to it that the sun was a part of daily life and ritual. In some cases, they humanized the sun so completely that it acquired definitely ungodlike traits. This Cherokee legend is not only entertaining, but demonstrates just how completely anthropomorphic the sun had become for a people who lived by its bounty, or could die from its excess.

"Now the Sun hated the people of the earth because they could never look at her without screwing up their faces. She said to her brother the Moon, 'My grandchildren are ugly, they grin all over their faces when they look at me.' But the Moon said, 'I like my younger brothers.' They always smiled pleasantly at him when they saw him in the sky because his rays were milder. The Sun was jealous and planned to kill all the

[*Continued on page 58*]

*Thou appearest beautifully on
     the horizon of heaven.
Thou living Aton, the beginning
     of life!
When thou art rising on the
     eastern horizon,
Thou has filled every land with
     thy beauty. . . .*

*Creator of seed in women.
Thou who makest fluid into
     men.
Who maintainest the son in the
     womb of his mother.*

*. . . When the chick speaks
     within the shell,
Thou givest him breath within
     it to maintain him . . .*

*The countries of Syria and
     Nubia, the land of Egypt.
Thou settest every man in his
     place.
Thou suppliest their
     necessities . . .*

Akhenaten, "Hymn to the Sun," translated by S.G.F. Brandon in *Religion in Ancient History*

Huichol Indian yarn paintings, *How One Person
Received a Huichol Name.*

How We Contemplate Hikuri in Wirikuta.

Where Offerings Are Made in the Sea.

Many native American myths show a belief that the sun was once scorchingly close to earth— surprisingly coincidental with the scientific hypothesis that the solar system resulted from the gradual changing of a nebula in which the sun was once closer to the earth.

According to the Cherokee, the sun first ran a track so close to the earth that the crayfish had his shell scorched bright red. Indian conjurers then put the sun another handbreadth higher in the sky, but it was still too hot. They raised it until it was seven handbreadths higher, just right in its arc, and here it remains in harmony with the earth's needs.

In late December, the Hopi Indians invoke the sun, asking it to turn back in its course, bringing new warmth and life to crops, animals, and people. With feasting and prayers for all creatures, the solstice itself is welcomed.

On a southwestern New Mexico butte, ancient American Indians built a stone solar instrument that measures the seasons. At summer solstice a "dagger of light" moves vertically down the slab bisecting the spiral stone carving.

people. So every day when she got near her daughter's house she sent down such sultry rays that there was a great fever and people died by the hundreds. They went for help to the Little Men who said the only way to save themselves was to kill the Sun. . . ."

Well, the Little Men did not kill the Sun, as all of us know. The Cherokee story winds on and on, charged with the imagery of many sun myths: the captive sun, the subservient sun, the sun that does not shine, and, finally, the dependable star of character, a star of heroic dimensions, a star quite different, and yet just as valid, as the star of the astrophysicists.

There are full-fledged sun traditions among the Eskimos, the ancient Greeks and Romans, the Egyptians, the Japanese, Africans, Mayans, Toltecs, Aztecs, Incas, South American tribes, and the entire spectrum of North American Indians, the Chinese, Hindu, Alaskans, New Zealanders, Polynesians, Scandinavians, Europeans, the Irish, and the French. It would be redundant to list every culture, every nation, every people, every island nation, every tribe that has woven the sun into the fabric of its being and its history. Somehow, in some fashion, the star has touched every group of humans gathered on every corner of this earth. And most of them have converted their myths to religion.

Among the deities declared to be the sun through the cultures of centuries past are: Saturn, Cronus, Jupiter, Pluto, Bacchus, Apollo, Janus, Pan, Hercules, Vulcan, Mercury, Osiris, Serapis, Baal, Adonis, Shamash, Moloch, Hammon, Bel-Samen, Asabinus, Ra, Atum, Ptah, Mandoo, Gom, Moni, Set, Mithras, Vishnu, Siva, Indru, Urotal, Belenus, Hu, Maui, Viracocha, Vitzliputzli and . . . the list could cover pages. And with such a number of sun-gods, it follows that there is also a vast collection of sun shrines, sun temples, and sun monuments: Stonehenge and the 80 other henges in England and Brittany, the pyramids and sun temples of the Incas still standing in Peru, an American Indian "woodhenge" in Illinois, the cliff markings of the Hopi, the monoliths of Easter Island, the

obelisks of the Mediterranean shores, the temples of Japan—everywhere there is witness to the serious business of building properly impressive structures where some sort of sun homage could be paid.

The sun is royal, the sun wears a crown, the sun has a halo, the sun is fire, the sun is a fiery chariot, the sun is a torch. The sun's rays, in their basic form, become the cross that, in turn, becomes the symbol of Christianity: a religion that gives Christ a crown of thorns and a halo; and, as if to make certain His relationship to the sun is not overlooked, moves the celebration of His birthday (which actually occurred in the spring) to the time of the winter solstice: the season (in the northern temperate zones) of renewal, of the sun's resurrection and that resurrec-

[*Continued on page 65*]

For centuries the Plains Indians, among them the Dakotas, Cree, Shoshoni, and Crow, have invoked the power of the sun by practicing the ritual sun dance. Moving back and forth from a central sacred pole to the lodge wall, the Indians danced to the accompaniment of music and singing for as many as three days and nights. In their dance they asked the sun-source for power, vision, success in their hunting or wars, and for the welfare of the tribe.

Today the Ute and Shoshoni tribes living on reservations still perform the dance to promote tribal unity and well-being. From it tribal holy men still gather power to cure the sick, control the rains, and see visions for their people.

According to the Indians of Mato Grosso, Brazil, the sun gave flutes to men and taught them the tunes to play and the dances to dance. When the cicadas begin singing and the rains make the rivers rise, men play the sacred flutes and dance as their ancestor, the sun, taught them to do.

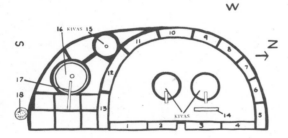

At their sun temple in Mesa Verde, Colorado, the Hopi Indians measured the striking of solar rays to time their religious rituals.

North American Indians often used sun symbols as part of their ceremonies as can be seen in these crown dance headdresses from the White River Apache.

Welcome, Great Chief, Father, as you come and show yourself this morning. Let nothing evil befall me on this day which you have fashioned according to your wishes, Great One, Walker-Across-the World, Chief!

Kwakiutl Indian prayer

*The sun is looking at me; it is
    looking down at me;
I, I am looking up at it.
I, I am happy, it is looking at
    me;
I, I am happy, I am looking at
    it.*

Navajo chant

*Don't you ever
You up in the sky
Don't you ever get tired
Of having the clouds between
    you and us?*

Nootka Indian song

Mexican religious painting.

61

From the Congo to the tundras of Hudson Bay, early peoples saw the sun as a great hunter whose dawn rays scattered the herds of the night sky, the stars. But just as the solar hunt never depleted the herd of stars, which came forth in glory again the following night, so early peoples measured their hunting by ritual so as not to deplete the antelope or other animals of the hunt. The African Pygmies made this solar connection clear by making a ritual kill of symbolic antelope at dawn just as the rays of the sun struck the pictured beast. Only after uniting their action by ritual to the movement of the universe did they go out to kill the real antelope.

Golden fern seed gathered at the mystic seasons of midsummer and winter solstice was thought by Bohemian peasants to be a special emanation of the sun which held the key to all the hidden treasures in the earth. If you placed a fern seed in your moneybag, they said, the money would never decrease, no matter how much of it you spent.

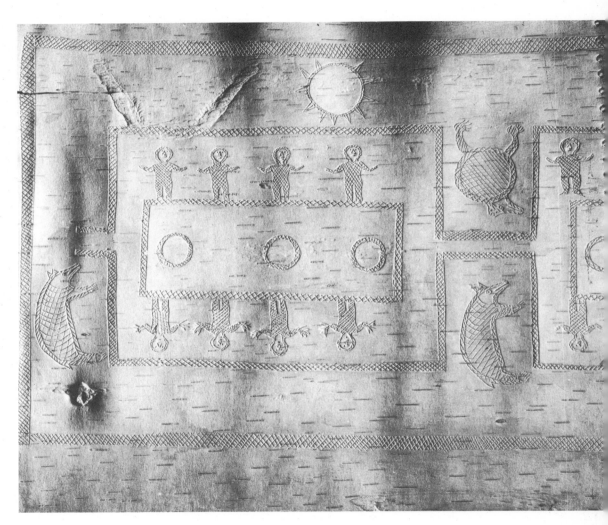

An Ojibwa bark scroll.

62

In Leyte, Sumatra, and other small islands near Australia, the people regard the sun as the male principle by which the earth becomes fertile. At the beginning of the rainy season, a ladder is placed in a holy fig tree to help the sun descend to earth. At a great feast to celebrate the union of sun and earth, men and women dance, feast, and join in symbolic sexual union beneath the tree.

In a myth from the valley of the Ganges River, Surya the sun was abandoned by his wife because his dazzling brightness made her tired. She took darkness into her bed in his place. Surya finally won back his wife by reducing his glitter to seven-eighths its original brightness, setting an example of the spirit of compromise that makes relationships possible.

Sunrise—A Gift No Gem Could Equal by Charles Lovato.

63

BIG SUR SUN SUTRA

Sun Sun
Ah sun Om sun
Sun Sun Sun
Great God Sun
Still riseth in our Rubaiyat
and strikes and strikes
And strikes the towers
with a shaft of light!
Strikes us in our Sun-ra boat
Strikes and strikes us in our
    Sun-ra boat
Sun Light Life Light
Sun Sun Sun
Great God Sun!

Lawrence Ferlinghetti, *Open Eye, Open Heart*

I praise the disk of the rising
    sun
red as a parrot's beak, sharp
    rayed,
friend of the lotus grove,
an earring for the goddess of the
    east.

Vidya (A.D. 700–1050) India

Pi disk from Chinese Han dynasty.

64

SUN FANCY

tion's implications of an afterlife, an immortality. Thus do the myths of the sun live with us still, even though we try to insist there is no place for myths in this, the Age of Science.

Yet we observe the solstices—summer and winter—and we take note of the vernal and autumnal equinoxes. And, in the United States, the first Americans—the Shoshones, the Utes, the Cree, the Dakota, and the Blackfeet—still dance their sun dances of purification and praise. We cannot, no matter how we try, dispose of 4,000 years of ceremony, tradition, superstition, religion, story, poem, and myth by repeating the facts of the sun's diameter, its core temperatures (Fahrenheit and Celsius), and its mileage from this earth. We are all still children of the sun, our star, and we need to know our father/mother in both his/her aspects—the factual and the fanciful—if we are to construct the most benign, the most beneficial relationship; if we are to draw our sustenance, our energy, from the sun overhead, instead of the smaller suns of fission and fusion that we can now create on earth. The Cherokee, the Eskimo, the Samoan, and the Arawa would tell us that to try to imitate the sun would be an affront, likely to make the sun in the sky angry. We should listen to the myth as well as the facts, for there is meaning in both.

It was myth that directed the architect of Saint Peter's Cathedral in Rome to orient that superb building so that it welcomes the spring on the morning of the vernal equinox. So exactly aligned to degrees east and west is the basilica that when the great doors are thrown open on the equinoctial dawn, the rising sun's rays reach through the outer doors, the inner doors, flash straight down the nave until, at last, they illuminate the altar.

And it is myth that directs the children who play "Oats, peas, beans, and barley grow. Here we go round the mulberry bush." As they do, they turn "sunwise"—clockwise, as we call it, from east to west—around the mulberry bush that is a symbol of the sun. We make our cakes stirring in the same direction as the sun moves, and we rub our bodies

65

India, seventeenth century. *Nadir al-Zaman*. An allegorical representation of Jahangir of India and Shah Abbas of Persia embracing in front of symbols of peace.

Christian tradition set the conception of the Christ Child at the time of the vernal equinox in spring and the birth of the child at the winter solstice. Commenting on the association of Christ with solar movement, Paulinus of Nola writes: For it is after the solstice, when Christ born in the flesh with the new sun transformed the season of cold winter, and, vouchsafing to mortal men a healing dawn, commanded the nights to decrease at his coming with advancing day.

With solar imagery, the Christian Church sings a welcome to the Christ Child in the evening antiphon of December 21:

O Day-spring, Brightness of the
        Light Eternal,
and Sun of Justice, come and
        enlighten those
who sit in darkness and the
        shadow of death.

The sunlike halos symbolize spiritual light in this twelfth-century painting *The Coronation of the Virgin.*

66

The association of Christ with solar light is symbolized in the golden monstrance, in which the consecrated host is held before believers during the service of benediction.

from left to right, from east to west, from sunrise to sunset. We are not as separate from the sun as any of us might think after reading a scientific description of the star's nuclear furnace. Historians and sociologists have gathered to take note of new discoveries of man-made sun markers in Vermont, Arizona, California, and Colorado. These are evidence that earlier Americans had learned what many of us seem to have forgotten: the sun is important to our daily lives. And as such, it should be observed, noted, and studied in each of its dimensions: astronomical, physical, and metaphysical. That is a lesson well worth heeding.

In the beginning God created the heavens and the earth. Now the earth was a formless void, there was darkness over the deep, and God's spirit hovered over the water.

God said, "Let there be light," and there was light. God saw that light was good . . . God made the two great lights: the greater light to govern the day and the smaller light to govern the night, and the stars . . . God saw that it was good.

Genesis 1:1–4; 16–18

A CONTEMPORARY WINTER SOLSTICE CELEBRATION

In a candlelit room, we sat together with friends talking about how we were at this turning point of the year, winter's longest night, the symbol of dark times in our lives. Quietly then, we took turns blowing out the candles, with each one mentioning some defeat, failure, or loss in our life during the past year, taking time to be with the pain of those moments.

When we had put out all but the last candle, one of us carried it out of the room, circling around the house before returning, himself a symbol of the returning new sun. As we relit the candles, again taking turns, we each spoke of a hope for the year to come. All candles again flaming, we sang together the simple melody: "Long is our winter, Dark is our night, Come set us free, O saving light."

Sara Ebenreck

Eskimo carved and painted shaman's mask.

As I write about the sun, I do so on the very day that it reaches its northernmost swing on its arc through our hemisphere. The summer solstice is occurring here in Maine on the cool, clear evening of June 21.

As I watch the western sky, quietly meditating on a celestial event I cannot see, I think of the ancients and their celebration of Midsummer Day—their slight misnomer for the time of the summer solstice. Those "pagans" (as they were later labeled by the Christian Church) celebrated the day with great festivity. Their exuberance was later toned down by the Church when, in the Middle Ages, it changed the day's name to Saint John's Day and used it to mark the birth of Saint John the Baptist. Throughout Europe great fires were built to honor Saint John's Eve—a blazing defense against emergence of the witches, warlocks, wraiths, and demons that were said to be especially active on this festival night—a night of Satan as opposed to a day of Light. It was this custom that Shakespeare had in mind when he wrote

Many children's games have been built around sun-legends. One, called "shove winter out," comes from the Indians of the islands at the southern tip of South America. In the game, the children are divided: half of them given a black mark on their forehead to indicate winter; the other half marked with green to symbolize summer. All the players fold their arms and keep them folded as they play. A large circle is drawn on the ground (or floor) and winter-children step inside. Using only shoulders and backs to shove and push, the summer-children try to push the winters out of the circle and occupy it themselves. Once winter-children are pushed outside, they must join the summers and help the shove against winter.

Solar lobby logo.

The relation of the powerful sun to the beauty of dawn is personified by the Greeks in the myth of Daphne, the dawn goddess, who springs from the waters at the first tint of morning light. Her fading colors in the growing light of day are a sign of Daphne's flight from the sun, says the myth, finally culminated by her transformation into a laurel tree.

In a Finnish myth, the sun is entrusted each evening into the hands of the maiden Evening Twilight who is to hide the fire so that no damage is done to it. At daybreak she is to give the sun to her brother Dawn who kindles it into new life. In summer when the Finnish night is brief, the two come together at midnight and their loving touch and glances turn the midnight skies red.

Chariots of the sun from Scandinavia.

A Midsummer Night's Dream, and that night which inspired the Russian composer Modest Mussorgsky to write *Night on Bald Mountain.*

It was the glyphs at Stonehenge, and the other henges, which marked the summer solstice for the Druids and other Celts; it is a brief notation in the almanacs which marks them for us. Most of us do not even

70

walk outdoors at the time of the solstice to try to sense the vast celestial event which is taking place some 93 million miles away. For most of us, such matters have become of no consequence. We have long since forgotten the myths of the Sumerians, the Greeks, the Celts, the Blackfeet, and the Cree; we no longer celebrate Midsummer Day or Midsummer Eve with festivals and bonfires. Instead, if we take note of the sun in any extraordinary ways, we do so by pausing a moment over reports of a debate between astronomers about the stability of the sun's dimensions.

If sun fancy as well as sun fact held a more significant place in the national scheme of things, citizens who are still adamantly skeptical

When the scientist steps back and begins to ask the big questions, he begins to think mythically. When the artist steps back and tries to address himself to the whole existence of man, he begins to think mythically. And there begins to be a convergence where they come together, so that if a physicist begins to talk about black holes in space and gravity-collapsed stars that are swallowing up everything and we begin with a big bang and we end with a black hole, in order even to describe these things he has to begin to speak in a mythological language . . .

William Irwin Thompson, "Planetary Culture and the New Image of Humanity"

Variations Within a Sphere Number 10, *The Sun,* by Richard Lippold. This sculpture is made of gold-filled wire in an abstract pattern of rays that shine like the sun itself.

The original Hawaiians say that years ago the sun used to burst up from the sea at dawn and race across the sky so quickly that people were cheated of their time for hunting and fishing. Then, they say, there arose a brave fisherman who wove long ropes to make a snare which he set over the waters at the very edge of the earth. Next morning, when the sun bolted straight into the snare, he scorched and blasted the fisherman, struggling for his freedom. But the fisherman held fast, balancing his canoe on the waters. At last the sun, seeing he had met his master, agreed to slow his pace. Having that promise the fisherman set him free, but left behind a few ropes to remind the sun of his promise. So, the Hawaiians say, we can see the ropes hanging down from the sun as it sets in the evening.

DONN P. CRANE

about the efficacy of energy from the sun could be more easily persuaded of the viability of a solar future. What science has stated in so many factual words is often not as believable as a fable Tecumseh might repeat to General Harrison. If we are to turn to the sun in our sky instead of attempting to concoct its miniature duplicates here on earth, we must begin to see it in each of its many dimensions. Of these, sun fact is just one, and, as we have seen, those "facts" are constantly changing. Sun fancy, on the other hand, is a solar aspect that needs to move from its modest, albeit encouraging, renaissance to the full-fledged restoration of sun myth, sun legend, and sun ritual. Certainly, there is room for both; one can complement the other. And when both are firmly in place in our consciousness, we can take the next step and begin to re-acquaint ourselves with the many ways the sun affects, sustains, and enhances this world we live in.

72

 *SUN FANCY*

For as regards light; since the sun is very beautiful with light and is as if the eye of the world, like a source of light or very brilliant torch, the sun illuminates, paints, and adorns the bodies of the rest of the world; the intermediate space is not itself light-giving, but light-filled and transparent and the channel through which light is conducted from its source, and there exist in this region the globes and the creatures upon which the light of the sun is poured and which make use of this light. The sphere of the fixed stars plays the role of the river-bed in which this river of light runs and is as it were [sic] an opaque and illuminated wall, reflecting and doubling the light of the sun: You have very properly likened it to a lantern, which shuts out the winds.

Thus in animals the cerebrum, the seat of the sensitive faculty imparts to the whole animal all its senses, and by the act of common sense causes the presence of all those senses as if arousing them and ordering them to keep watch. And in another way, in this simile, the sun is the image of common sense; the globes in the language . . . intermediate space of the sense-organs; and the sphere of the fixed stars of the sensible objects.

Johannes Kepler, "Epitome of Copernican Astronomy"

The sun, it shines everywhere.
                    SHAKESPEARE

3   *SUN NATURAL*

"As the days commence to lengthen, the cold begins to strengthen." This is a Yankee saying that underscores a natural fact in the upper latitudes of both hemispheres: as the earth tips on its axis and the winter solstice marks the beginning of the sun's return to a more direct, overhead angle, the coldest days of the season are still to be endured. The paradox of longer hours of sunshine and colder readings on the thermometer is created by the earth's heat-gain/heat-loss rhythm. Like a fire in a parlor stove that takes hours to fully warm a farmhouse on a December morning, the sun needs time to restore an earth that has—in the Northern Hemisphere—been losing heat since early September.

For millions of Yankees, the earth finally begins to shake winter's grip when a series of natural—not astronomical—events takes place, events like the rising of sap in sugar maple trees, the northward migration of wild geese, the fluttering of kites in a March sky.

But of each of the hundreds of seasonal signs, vivid or subtle, commonplace or rare, none is more widely heralded or more eagerly anticipated than ice-out. Like drought-stricken farmers whose newly planted crops will fail unless the rains descend, the winter-weary people of the entire northern tier of states search the sky and the vast frozen surfaces of lakes, rivers, ponds, and bays for a sign that the icy armor is weakening. In village stores bets are made and recorded as normally prudent folk wager their hard-earned money on their estimates of the hour and the day that the bay or river will run free. They do it more as a kind of sacrifice to the gods of spring than as a practical gamble; the logic holds that if enough wealth is at stake, the forces that keep the ice in place will somehow be weakened.

We are wont to forget that the sun looks on our cultivated fields and on the prairies and forests without distinction. They all reflect and absorb his rays alike, and the former make but a small part of the glorious picture which he beholds in his daily course. In his view the earth is all equally cultivated like a garden.
. . . We might try our lives by a thousand simple tests; as, for instance, that the same sun which ripens my beans illumines at one a system of earths like ours.
Henry David Thoreau, *Walden*

Twilight in the Wilderness by Frederic Church.

Yet there is no mystery in the elemental forces at work. On the one hand, there is the span and intensity of the waning winter to be considered. For each day temperatures cringe below the freezing mark, more molecules of water are transformed from liquid to solid, more micrometers of ice are added to the sheets already in place. Where I live in Maine, some winters have created great masses of bay ice more than two feet thick—a layer so monumental, so overwhelming that it subdues every memory of the water that once frolicked in the space that has

Winter is more than a season
bounded by a solstice and an
equinox, more even than
snowstorms and icebound lakes and
wind roaring down from the Arctic
tundra. It is primitive forces at
work, cleansing and clarifying the
earth. But it is also beautiful and
awesome and full of wonder.
Hal Borland, *Seasons*

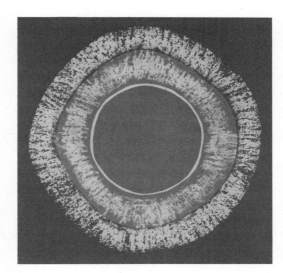

Although the sun daily shoots up high in the sky, for the equatorial regions of the world there also are seasons, marked less by day length or cold than by the coming and going of rains. For the equatorial zone, the peak rainy times are April and November, both followed by drier seasons, but there is no month without rain. In a rainforest where growth happens year-round, trees do not show the clear, yearly growth-ring markings of trees in a temperate zone. Food can be grown in all seasons.

Further north in India and Pakistan, the drought and rain are more extreme, producing the climate of the monsoon. In May India is dry and hot. Suddenly water-laden air from the oceans bursts forth and the June monsoon season begins—to last until September when it abruptly ends, leaving October and November dry. With too little or too much rain, the rice crops fail, making the monsoons all-powerful in the lives of the people.

become a sterile, pale, and frigid expanse. Pressed from shore to shore, reaching as far as the eye can scan toward the seaward horizon, the vast deserts of ice weigh as heavily on the soul as they do on the waters they hold prisoner. "If only we could see the blue water dance in the sun," the winter-weary tell themselves, "then we could believe that spring will come again."

It is testimony to the pure intensity of northern winters that we lose faith in the sun, that we scrutinize the ice of our bays, lakes, and rivers as if we are convinced that this year it will not surrender, that this year our bright waters will not be released. Instead of marking the length of the ice seams that span the channels, we should be looking overhead at the sun that surges toward the solstice. For as surely as the sap rises in the maples, that sun is gathering its warmth in the incremental degrees of its ascent in northern skies. And, just as certainly, on the clear days of March, nearly half our star's radiant energy is transformed to heat when it touches the white reaches of our winter rivers.

When, at last, we see a sign of weakness—a crack that lengthens as we watch, a groan that rumbles from below, a jut of tormented ice pushed skyward by impatient tides—we imagine it is the waters beneath that surge to their victory. We all but ignore the sun overhead. It is, we say, the water's triumph when cracks become chasms and icy fortresses are breached, one after the other. Incredibly, the great horizontal walls crumble, booming as they move, pushed by March winds, tugged by spring tides, tumbled on surging freshets. Almost overnight, it seems, the victory is won. Winter is dispatched.

To watch as winter's ice is vanquished is one of the most dramatic of seasonal moments; and it is these moments, ironically, which tend to overshadow the timeless celestial pageant that has been played century after century for tens of thousands of centuries. That drama, seen through the prism of time compressed, has one hero, one protagonist— the sun. And it is still the sun, even though we so often fail to recognize its

many dimensions, that influences, directs, causes, shapes, sculpts, creates, and designs the diverse patterns of our environment.

It matters not whether you are on the deserts of Arizona where the rising sun shatters the stillness of freezing nights with bludgeons of heat that raise afternoon temperatures to more than 100°F., or whether you are a resident of downtown Miami who is sun-warmed during January, but sun-cowed during June when the sun's tropical intensity forces nearly everyone indoors for the respite sheltered air conditioning brings. Wherever there are people, from the equator to the poles, the sun is an overwhelming presence, either by its absence during the winters of New England, or by its dominance of a Floridian's July.

We could understand our sun's prevailing influence more clearly, perhaps, if we look at it through that prism of time compressed instead of watching it set on a March evening as ice floes move down the bay. Instead of the days in March, split as they are by the bettors at the general store into the hours and minutes of "ice-out," consider the centuries in a millennium when the sun waged war with sheets of ice that covered continents instead of the Great Lakes, that imprisoned a hemisphere instead of Hudson Bay.

There have been a series of such chill dramas in the planet's history; the one most often signified by the term Ice Age began to end some 10,000 years ago. Until then, the Northern Hemisphere's nations, which now take their climates quite for granted, were imprisoned by vast sheets of ice. These glaciers had crept southward for centuries, grinding down mountains, obliterating riverbeds, and burying green forests beneath tons of frozen seawater, evaporated from the oceans by the sun, and returned as snow from the skies to an earth made frigid by some elemental shift in solar cycles. Then, instead of watching and waiting through the lengthening days of March for their river, lake, or bay to be released from winter's grip, the people of North America, Ireland, Great Britain, Greenland, Scandinavia, Russia, and other northern nations watched as

Ice Age glaciers reached well into areas now known as fertile farmland.

The Wolverine Glacier near
Seward, Alaska.

their farms were buried under the steadily advancing ice.

If this happened today, the ice-out that would follow as the glaciers retreated would be infinitely more important to the planet and its people than the ice-outs we Yankees take for granted every spring. But the same elemental influences that cause one bring about the other and, of these, none is more influential than the sun. Functioning in ways that are not yet fully understood, the sun is primarily responsible for our climate, both annually and through the centuries.

While scientific debate still surges as to the detailed causes of the Ice Age that ended some 10,000 years ago, and the echo of that bleak time known as the Little Ice Age of the seventeenth century, scientists who have studied the warm–cold cycles of the millenia theorize that a variety of events on the surface of the sun—sun storms, sunspots, solar flares—and events on earth combined to lower temperatures enough to create giant glaciers where cattle once grazed. The earth's variables that were the major influence on starting, and ending, the Ice Age are primarily involved with its motions in space. This globe of ours not only rotates on its axis (once every 24 hours) but it wobbles like a spinning top from side to side. That wobble takes about 10,000 years to tip the earth's axis in one direction, and another 10,000 to recover and lean to the other side. In each cycle, more—or less—of one hemisphere or the other is exposed—or denied—a portion of sunlight. Currently, give or take a century or two, the Northern Hemisphere is in its darkest time of the full, 21,000-year cycle, a circumstance which has helped increase talk that a second ice age is on its way.

Most scientists agree, however, that it takes more than the "precession of equinoxes" (the astronomer's term for the 10,000-year wobble) to awaken the somnolent glaciers and get them moving like a herd of huge, white mammoths migrating south, burying everything as they go. Sunspots are a popular phenomenon for discussion in any ice-age debate. These sites of enormous solar storms—unlike anything we can

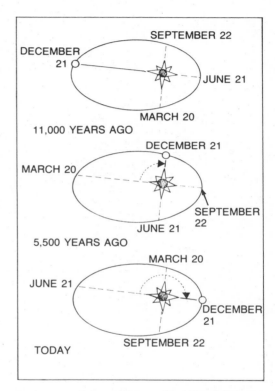

Precession of the equinoxes. Because of the combined westward motion along the elliptic earth orbit and planetary forces which slightly alter the plane of the earth's orbit, the positions of equinox and solstice shift slowly in a cycle of 22,000 years. The drawing shows winter solstice occurring at three points of that cycle. As a result, the distance from the sun to the earth on solstices and equinoxes changes yearly.

Because of gravitational pull from the sun and moon on the equatorial region of the earth, its axis of rotation changes slowly completing a cycle every 26,000 years.

imagine—produce great magnetic vortexes in the sun's hot, electrified gas.

The solar eruptions also have been cited as the cause of weather changes, variations in radio and television reception, and influences on the advance, or retreat, of the polar ice caps. But the sun has other solar flares which are quite different from its sunspot maelstroms; they also have an effect on our climate and our long-term weather patterns, and again, there is much debate on the precise dimensions of that effect.

In addition, there are the sun's cosmic rays—the invisible "atomic bullets" that bombard the earth. The volume upon arrival here increases in relation to increases in the activity of the sun's magnetic fields. Beyond these three, most frequently cited solar events which affect earth's climate and the ponderous rhythms of its ice-ins and ice-outs, there is the flickering of the sun itself, the appearance and disappearance of dust in the earth's atmosphere, and the overall increase of carbon dioxide in that atmospheric layer. This increase has been caused by technology and its fires of coal, oil, and natural gas, which, in addition to powering the Industrial Age, have added enough carbon dioxide to the scheme of things to alter the heat-exchange patterns of the planet.

We are, some observers will argue, burning so much fossil fuel that the "greenhouse effect" of our increasing carbon-dioxide layer will not only alter the rhythms of the expansion and contraction of the polar ice caps, but may well raise global temperatures enough to melt most of the polar ice. Then, instead of being crushed by advancing superglaciers, a city like Boston will be flooded by the rising waters of the Atlantic, swelled as it is with the floods of a melting Arctic.

You can take your pick of the various cosmic, natural, and man-made influences that will, or perhaps already have, determined the scale of the next ice age, or, indeed, whether there will be one. Some of those influences have yet to be listed here. There are, for example, the jet

streams: those high winds that blow at velocities of more than 100 miles per hour far above the earth. They are affected by the sun, as well as the earth's magnetic field and its variations. Together with the increasing influence of man-made dust in the atmosphere, the jet streams may also determine whether we have glaciers or grapefruit in our backyards.

Dust has influenced the weather in the past, and, through the weather, the people who lived with it. Periods of below-normal temperatures and damp weather have generally followed in the wake of major volcanic eruptions. The popular scientific theory suggests that dust from volcanoes helps to weaken the sun's heat, allowing the formation of more clouds and consequent precipitation. Science writer and weather specialist Nigel Calder (*The Weather Machine* and other books) theorizes that one of the most popular monsters ever to walk the imaginations of millions of humans was a child of a volcanic dust storm. The dust veil that followed the 1815 eruption of the Indonesian volcano Tambora, he tells us, caused an extremely wet and foreboding summer in Switzerland the following year. To help ease the burden of forced days indoors, Lord Byron suggested to Mary Shelley, his houseguest at his Swiss chalet, that she try writing a story. Evidently, her gloomy thoughts matched the weather, for she gave the world the tale of *Frankenstein*.

According to some prognosticators, another generation of Frankensteins may be on the way. They argue that the amount of man-made dust has now reached proportions that surpass the volcanic aftermaths of more than a century ago. For them, the future of the polar ice cap is assured as more and more dust fills the air in the wake of massive earth-moving efforts, construction, mining, and agriculture dependent on chemicals that prevent the soil from maintaining its organic abilities to retain moisture.

What are the weather patterns of the coming century? When will the next continental glacier reach Detroit, Michigan, or Dublin, Ireland? Before anyone can answer, they must consider the effects of dust in the atmosphere, carbon dioxide in the stratosphere, cosmic rays, jet

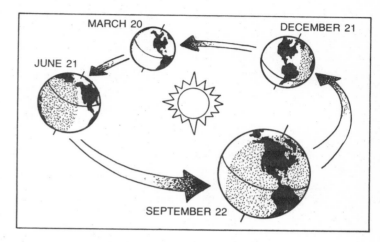

The seasonal movement of the tilted earth.

Give me the splendid silent sun
with all his beams
full-dazzling,
Give me juicy autumnal fruit
ripe and red from the
orchard,
Give me a field where the
unmow'd grass grows . . .
Give me solitude, give me
Nature, give me again O
Nature your primal
sanities!

Walt Whitman, "Give Me the Splendid Silent Sun"

If you put a stake through the snow on a sunny December 21 and mark the angle and length of the shadow at, say 2:00 P.M., the following year the same shadow will fall at the same place at the same time. It will not, however, be quite the same on any of the other 364 days. In the temperate zones we are in motion in a more recognizable way than the inhabitants of the equatorial regions. As the earth tips so the maximum amount of sun strikes the "upper" half (the Northern Hemisphere), we know June and the splendor of summer. One hundred and eighty days later, halfway around its circling of the sun, the North Pole is tipped its farthest from the sun; New

[*Continued on page 87*]

streams, pollution, volcanic eruptions, the sun's magnetic fields, the sun's stability, sun storms, sunspots, sun flares, and the 10,000-year precession of the equinoxes—to cite just a few of the factors that any prudent forecaster should take into account.

It is safe to say, in my opinion, that a list of options of such volatility and unpredictability will eliminate every "prudent" prognosticator from the field. There are, and there will continue to be, too many unknowns, too many mysteries still shrouded in the flames of the star we call our sun to be able to predict with certainty whether Boston will be buried in ice, or flooded by the waters of a melting Arctic. Indeed, the chances are equally good, or better, that Boston will stay as bustling as it is—a place of "typical" New England weather: cold and raw in winter, windy and wonderful in the spring, warm and even warmer in the summer, and invigorating in the fall.

And it is the sun and its relationship to the earth that creates this splendor of the seasons. It is the regular, and yet irregular, wonder of seasonal change that kindles the brightest creative fires. Yet, in a staggering irony, as we have progressed along the road of discovery and technological invention, we have done our best to deny those seasons, to create our own, controlled environments shaded from the sun, sheltered from the rain, protected from the snow, and barricaded by stone and concrete walls that prevent the winds from caressing our cheeks or bringing us the smell of new-mown hay.

We would do well to get re-acquainted with the seasons. They are a gift of the sun, bestowed as a reward for the delightfully eccentric pattern of the motions of this planet.

In the northeastern United States, for example, the sun is farthest north (or the earth is tipped its farthest "down" on its axis) within a few hours of June 22—the date that marks the solstice, the official start of summer, and the date that gives residents of the northern temperate zone their longest day of the year. In the Southern Hemisphere,

however, Argentinians and South Africans are sleeping through their longest winter night. By September 23, the autumnal equinox, the angle of the planet has shifted halfway toward its southern extreme; both hemispheres now have—for this one day—equal amounts of sunlight and darkness. Then the slide begins for northerners toward the winter solstice—the day of the longest night, but also the day in late December (usually the 22nd) when the sun begins its return journey; or, to state

[*Continued on page 90*]

Englanders are deep in winter, while the people of Tierra del Fuego are enjoying their longest day of the year.

The earth's wobble, its roll, and the "stretch" of its orbit around the sun alter the angle of the star's relationships to aspects of this planet. But, if you wait 21,000 years, or 700,000, depending on which cycle needs to be completed and begun again, this positioning of the two celestial bodies will repeat. This is the primary trait of the sun-earth friendship: they are circular, round, and free-floating in the vast reaches of space; their orderly movements are each cyclical—they end, sooner or later, where they began.

THE TRUE SOLAR YEAR

As if Christmas expenses, on top of inflation and recession and fuel shortages, weren't enough to worry us, now we have winter. In a few days, the weatherman's going to remind us that it is officially here. December 21st, isn't it? Or is it the 22nd? I can never remember. Anyway, each year, when the first day of winter is announced, we're all supposed to groan in unison—and why not, with the prospect of three months till spring ahead of us? So we all groan our groans at the news and then turn back to the job of getting through the rest of the holidays.

Bah!

Have I got you in a really bad mood? Have I painted the holiday picture a convincing shade of black? Well, surprise! There's good news. I just wanted to set you up for it.

The good news is called prosperity, and it's all related to the day, later this week, that we here in America call the first day of winter. Why we call it that I can't imagine, because December 21st—or 22nd—is

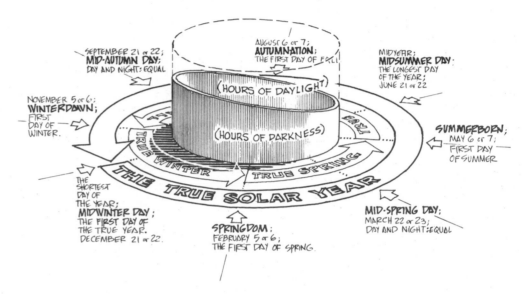

The true solar year.

the middle day of winter. It really is. You may not have noticed, but the true winter season arrived without any fanfare around election day last month. So, later this week, we'll be greeting Midwinter Day, which happens also to be the first day of the true year.

It's the day the sun starts its long trip north again. Even though three cold months still lie ahead, the days

will begin getting longer and sunnier by the end of this week.

Happy True Year!

Here's a true festival day, and we haven't even been told about it. Those idiots at the weather bureau, who know better, resist any kind of change. But all over the Northern Hemisphere, from Stockholm to Fairbanks, sun-people are beginning to smile. They know what's what on

earth. They know how important the sun is, and they revel in its return.

Midwinter Day requires no gifts, no turkey dinners, and no office parties. All you have to do is walk outdoors and smile, and the warmth will spread. You'll start to see an always-before-unnoticed brightness in the season, and, as the weeks slip by toward early February (when True Spring begins) you'll see the whole nature-world getting brighter, too.

At this point you're supposed to be wondering what this all has to do with prosperity. Well, the connection may seem farfetched at first, but it's as solid and dependable as the earth's movements.

Obviously, most of our economic troubles have been caused by a refusal, on the part of those in power, to face the facts of life. School children know them but the old pols sure don't. Kids know all resources are limited, that they must be carefully recycled forever.

Modern history has seen a brief spurt of supergrowth in no way resembling the normal state of things on earth. Now we're paying the price: inflation, unemployment, and shortages of all kinds.

In this Twentieth Century spree of ours we've managed to flush, dump, erode, lose, or incinerate enough precious materials to keep a sane society going forever. Many of those materials now lie, unrecoverable, at the bottom of the sea. Others just fell into cracks or blew away into the sky, and here we are on another Christmas binge, knowing it's all wrong, but going through the motions because we know no other way.

History has a huge gong and some flashing lights that go off each time mankind enters another great era. They went off when we stopped being nomads and invented farming. Then they gathered dust for a while—for thousands of years, in fact—until the age of science came along. Then they began to ring at closer and closer intervals as we swept, like spoiled children, through the ages of medicine, electricity, communications, electronics, the bomb, data processing, space flight, and now . . . what?

Now we've got to grow up.

We live in an engine, called life, that runs entirely on sun-power. Every bit of food we eat, air we breathe, fuel we burn—wind, rain, cockroaches, movie stars—all were made on the sun-powered machine. Still, we talk about solar energy as if it were one of many choices open to us when in fact everything around us, and we ourselves, run on nothing but sunlight. (Unplug that machine and see what happens!)

Now let's go back to Midwinter Day. It's the key, in a way, to whether or not we're going to pull out of our mess. As long as we say, "Oh yeah, Midwinter Day; so what?" we'll be heading straight for famine, terror, and nuclear disaster. But the minute we get down on our knees and bless that traveling old sun, the gong will gong and the lights will flash, for we will have entered the era of prosperity, when science will serve rather than thwart natural processes, and our economy will be built on the only kind of wealth that never goes bankrupt: sunlight.

Happy True Year!

The American golden plover summers in the Arctic tundra and winters in South America.

it more accurately, when the earth's journey around the sun reaches the point at which the North Pole begins to angle toward the sun instead of away from it.

These physics of the seasons are described so often in every general science schoolbook that the intricate, eternal movements of the spheres are most often considered commonplace. We are reminded of the celestial rhythms by matters as mundane as the arrival, or departure, of daylight saving time, the first Christmas carol—now scheduled for the day after Thanksgiving—or a January white sale. For most of us, the Internal Revenue Service personal-income-tax deadline on April 15 is more of a sign of spring than the March solstice.

Perhaps, someday, as we become more fully re-acquainted with the sun, we will set aside each solstice and equinox for observances as significant as those accorded any national, or international, event. There is more than baseball and the end of the school year to spring, more than ice skating and holiday shopping to winter. For every creature of the planet's temperate zones, except humans, the interrelated movements of the sun and the earth are the primal presence, the conductor's baton to which each instrument of their lives is orchestrated. It is neither wise nor realistic for us to assume that because we have invented the light bulb we can ignore those movements. Instead of thinking of September as a time when schools open after the summer holiday, we would do well to observe the monarch butterfly, the striped bass, the black-bellied plover, the polar bear, the Siberian crane, the wildebeest, the humpback whale, the swan, and the swallow.

For these creatures, and for tens of thousands of other species of the world's insect, animal, and marine inhabitants, the procession of the seasons, the lengthening of shadows, the passage of the solstice, the arrival of the equinox, are signals as demanding as a factory whistle. All around us, the other living beings of the planet are stirring, embarking or disembarking from journeys begun and ended and begun again by the

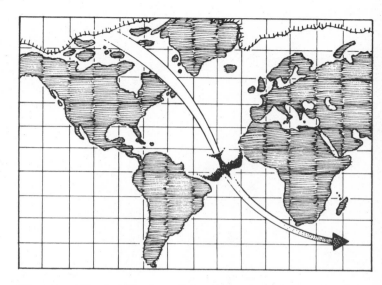

The extraordinary migration of the Arctic tern from near the North Pole to the Cape of Good Hope is guided by an internal sun-compass which determines direction by reference to the sun.

position of the sun in the sky. Guided by inner compasses which no scientist has yet fully deciphered, compelled by eons of migratory compulsions, the birds of the air, the fish and mammals of the sea, the insects of that largest kingdom, and animals from Africa to the Arctic move precisely with the declinations of the sun, crossing national and continental boundaries on maps in delightfully innocent ignorance of such invisible realities, determined only to do homage to solar rhythms we have all but forgotten.

In the Septembers of my boyhood, I would watch from an eastern Long Island salt marsh as plover massed for their seasonal flight. In huge flocks the shorebirds would rise, form groups, fragment as they were airborne, then regroup again, then rest, then rise, restless to be off on their 1,500-mile journey to the northern coast of South America, but waiting for the soundless signal, some subtle gradient of sunlight, some longer moment of twlight and dawn, that would send them on their way. The birds measure a critical span of daylight and dark with internal calibrators as precise as any engineer's.

Hawks sweep on the northwest winds of those Septembers of my memories. Riding the rapids of the air, they make their winged journeys from Maine to Pennsylvania and beyond. For them, for me, the equinox is a time of stirrings, in the skies above my horizons and in the seas beneath them.

Off Montauk Point—that stony and windswept finger of land that points forever toward the open Atlantic—a million fish gather for their autumnal travels. Sensitive to changes of water temperature calibrated in fractions of degrees, the striped bass, bluefish, albacore, and other quicksilver inhabitants of the temperate ocean tumble in the setting sun of a late September afternoon. As they sense the first cooling of their world, as they measure the minute increments of darkness, they exhibit a turbulent, almost reckless energy. Slashing through gathering schools of baitfish, they feed to store fat for their journey, now just weeks away. In

LXXIII

I'll tell you how the sun rose,–
A ribbon at a time.
The steeples swam in amethyst,
The news like squirrels ran.

The hills untied their bonnets,
The bobolinks begun.
Then I said softly to myself,
"That must have been the sun!"
Emily Dickinson

Our planet earth knows seasons that differ as radically as Arctic ice from jungle growth. Its temperatures range from −127°F. in Antarctica to

[Continued on page 93]

their frenzy, they attract clouds of herring gulls, black back gulls, terns, and gannets that whirl and dive in feathered blizzards that mirror the white water tossed by teeming fish.

That same seasonal activity extends far beyond the horizons of my memories. Over the rooftops of Oslo and Berlin, the European swallow takes wing on a migratory flight that will end only when the small, seemingly fragile bird reaches central Africa, more than 6,000 miles to the south. Spanning continents, nations, cultures, and seas, the determined bird's passage is fraught with international ironies. Protected in the northern nations where it spends its summer, the swallow flies to places where it is netted by the thousands for food and feathers.

From Siberia's vastness, the Bewick's swan and the Siberian crane are as restless as the swallow, the plover, and the bluefish. The swans will soon fly 2,600 miles from the Yamal Peninsula to the Irish marshes; the 350 Siberian cranes still surviving on the planet will soar across Afghanistan wilderness, dropping down for a spell to rest and feed, and then continuing to their wintering ground at Deoladeo Ghana in northern India.

Meanwhile, at the edge of the Arctic Circle, along the rim of the Bering Sea, polar bears pad across the top of the world. Shambling over glaciers, swimming in frigid waters 20 miles or more at a time, the bears travel 1,000 miles on migrations unseen by all but a handful of the globe's 4 billion humans. The bear follows the seal, who follows the tenuous warmth of the erratic Arctic sun. In an echo at the opposite end of the earth, the first of the humpback whales arrive to feed on plankton blooms in the Antarctic Ocean—blooms spawned by the same oblique rays of a sun that touch both poles with its longest reach. The 50-ton whales have cruised from Hawaii, sometimes leaping clear of the Pacific in the exuberance of their adventure, then crashing back with monumental splashes that send showers of spray into the sunlight.

We can also witness another kind of migration—this one a

journey without movement. While the plover flies from Montauk across the equator to South America, the deciduous trees of the northern temperate zone are undertaking a different sort of seasonal activity. As the September sun's shadows lengthen, allowing frost its first foothold of the fall, sugar is trapped in the leaves of maple, birch, aspen, oak, and the other hardwoods of the North. For travelers, the event is a tapestry to be viewed from the seats of the family car, cruising through valleys on a foliage tour. The tourists are rewarded with one of the season's most vivid spectacles, yet many of them see only the surface events and drive on, quite unaware that they are witnessing nothing less than the metamorphosis of a process that is the fundamental key to life itself.

The nature of that process is instinctively understood by the feathered, finned, and four-footed migrants who move with the turning of the leaves. They admit to being creatures of the sun and properly acknowledge its pivotal influence on their lives.

What is happening when leaves turn color on northern hillsides is nothing less than the shutting down of life-support systems in one temperate zone and the transfer of operations to the other. As the sun and the earth move through the steps of their measured and timeless dance, the light that is strong enough to power photosynthesis in the upper latitudes of one hemisphere is moved across the equator to begin awakening leaves that have been dormant during the star's absence. This motionless migration, this most crucial of any of the earth's processes, does more than color leaves on a Vermont mountainside; it transfers the energy of life from one part of the globe to another, creating, as it does, a balanced mechanism that insures the stability of the planet's atmosphere and the production of its food.

What we are seeing when we marvel at the splendor of autumnal foliage is nothing less than evidence of the sun's command of our mortality, for we are, all of us, creatures of the sun. Like the wild migrants who continue to acknowledge this as they travel thousands of

a summer-shade temperature of 136°F. in Mexico; its winds from hurricanes to none at all.

January in the high Arctic is a season without sunshine except for a warm glow seen at midday. Birds are rare then, animals mostly in winter sleep. At month's end the light glow is sufficient to allow seeing the color of an animal.

Arctic April sees the snow bunting arrive and the raven start to lay eggs. By its end, the sun–night relationship has reversed with almost 24 hours of light bringing the danger of snow blindness. The ground squirrel wakes up and animals mate.

By June, Arctic flowers are bursting out, birds' eggs hatching, and insects swarming. Eskimos move from their winter ice igloos to tents. But the summer warmth is brief. By September, killing frost is back, lakes freeze, and snow reappears. Animals shift their colors to winter white while Eskimos make their fur pelts into winter clothes. The sun again sinks low, reducing daylight by a half hour each day. The year will end in silent darkness.

As green plants eat radiant energy they make it into the chemical bonds of sugar molecules.

Sunlight which is not used in photosynthesis is absorbed into the ground and into the vegetation of a forest. Without this heat trees could not survive. As part of another cycle, the sun evaporates the water which the tree has brought up through its roots, trunk, and branches to its leaves. Each year the transpiration process moves a 20-inch layer of water without eroding the landscape.

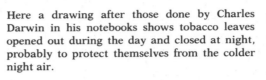

Here a drawing after those done by Charles Darwin in his notebooks shows tobacco leaves opened out during the day and closed at night, probably to protect themselves from the colder night air.

risky miles to follow their star, we are dependent on that star's life-giving mechanisms—delicate and subtle mechanisms that are as fragile as a cloverleaf, as delicate as a rose.

How often we seem to forget this simple truth. We argue the merits of solar energy as if the term signifies an exotic science so removed from our ken that centuries may be needed to decipher its intricacies. But for more than a century we have known that solar energy powers life on this planet, and, in recent decades, we have learned something about how this happens.

We have not, however, reduced the complexities of photosynthesis (putting together with light) to first-grade reading. There are unknowns, mysteries, wonders, even miracles if we are to consider photosynthesis within the context of other ages, other cultures, other times. We know this much, according to our textbooks: the energy that is born so tumultuously deep within the sun's seething core, that by-product of a fusion so vast and so violent that it is beyond our imagining, that explosion of heat and light that bathes our entire solar system is, by the time a minute fraction of it reaches earth, an energy that green plants can utilize. Left to survive in sunlight, we, the wildebeest, the plover, and the

A blazing sun beats in rhythm with the fields in this painting by Vincent Van Gogh entitled *Walled Field (Seen from the Asylum at St. Remy) 1889–1890.*

Humans get energy indirectly from the sun as it is passed through plants or animals and fish to us.

As long as the sun shines and the food-energy paths remain flowing, life-energy is constantly restored to our planet by the green plants.

Quietly, largely unseen by humans, the sun's light and heat power a salty wheel of life in the oceans. In the temperate zone, as the earth turns away from the sun, cooling waters send plankton into the rest of winter hibernation. Spring's strong solar rays bring an outburst of plant growth as tides and currents, themselves pumped by the sun, circulate fertilizers through the water. Summer's warmth and calm bring animal plankton to graze on the plants until autumn's winds again bring a mixing of sea layers, rather like a plowing of the waters.

swan would soon perish. But the vegetation of the planet can feed on the sun, using its rays to convert the carbon dioxide in the air and water from the ground to sugar and oxygen, and would survive.

The great cycle, the conversion of solar energy to chemical energy, could not be accomplished without the plants. Together with the animals which, in turn, convert plant energy to yet other forms that are utilized by humans, the plants constitute the planet's biomass—the total collection of living organisms. But as vast as it seems to be, that biomass is limited by the amount of sunlight that makes its way across space—a continuous supply of energy equal to about 2½ billion horsepower (as the engineers might evaluate it). Much of this solar energy is reflected by clouds, soaked up by the earth itself, used in the evaporation process, absorbed to power the winds, and warm the land and water. Just four-hundredths of 1 percent is utilized in the photosynthesis process—the process that supplies us with life itself, that gives us our sustenance.

Point zero four, four-hundredths of 1 percent. The earth catches two-one-billionths of the sun's power; of that, four-hundredths of 1 percent sustains life, maintains the balance of the atmosphere, recycles carbon dioxide, produces oxygen, grows corn in the fields, and not only gives us the energy to drive a carful of relatives through the Green Mountains of Vermont in October, but provides the energy to power that car with petroleum products refined from plant and animal biomass that decayed in millenia past. And we have the temerity to argue the practicality of solar energy? Well, some of us do. Some of us argue for our own fission and fusion suns, made by us here on earth because—we are told—such devices are more practical, cost less, and can be developed more quickly. How would those arguments stack up when balanced against the sun's generosity, the sun's lavish squandering of its energy in space—energy that may one day be re-directed to earth by a civilization that may, incredibly, grow enough to utilize all of the continuous supply of 2½ billion horsepower the sun sends this way.

Sunset over the Seine by Claude Monet.

The skies between the earth and distant sun in the direction of prevailing winds (which for the northern temperate climate means the western sky) are first to show the signs of approaching weather.

"Gray evening sky, not one day dry" is the weather wisdom based on the fact that gray skies mean that western air-borne dust is heavy with water droplets that will bring rain your way. A deep red sky, produced by damp dust particles, may also spill rain the next day, but a pink radiance means lots of dry dust in the air and a strong probability of fair weather coming. "Red sky at night, sailor's delight" actually means pale red or pink in the western sky. Blues and purples are a sign of medium dust and vapor concentrations, probably leading to continued fair weather if they dominate the sunset sky.

The way we "see" things often has more to do with what we cannot see than what we can. We have discovered, for example, that the incidence of random violence in many American cities continues to rise, in spite of the increments of increased wealth shared by larger and larger percentages of urban populations. This is not to imply that the war against poverty has been won. It has not. But the acts of violence against property and people are not directly related to income levels; they are, instead, much more closely related to densities of urban populations, and to a particular kind of urban environment.

A good argument could be made that random violence is so prevalent in our cities because those cities are so bereft of natural presences. Surely, that's one of the reasons. Acres of concrete, miles of asphalt, windowless buildings, and "controlled" climates have forced a "cold-turkey" denial of natural presences on tens of millions of people whose forebears had lived with nature for 10,000 years. It is possible, is it not, that this psychic trauma could trigger violence acts of aggression? One way to find out would be to convert cities to place where the sun, the wind, the stars, and the rain are restored in their elemental purity. Instead of centralized parks that provide city dwellers with their only opportunity to touch a blade of grass, the buildings themselves could be designed as green spaces, filled with patches of honest soil, with room for crops and creatures. Then at least, the 90 percent of the population that's squeezed into 10 percent of the space would be conscious of the weather, the equinox, the solstice, and other solar events.

To be conscious of the weather can be the beginning of understanding it; to be unconscious of it is to eliminate any chance of gaining that understanding, and also any chance of a closer acquaintance with the sun. Just as it sets the clocks for global migrations, fuels the cycle of life with photosynthesis, establishes the dimensions of the ice caps, and inspires the author of *Frankenstein*, the sun is the primary constructor of the weather patterns that affect the lives of every inhabitant of the planet.

 *SUN NATURAL*

Given the universal significance of those patterns, it is, at the very least, risky to suggest that we can live in utter ignorance of them.

Without the sun, we would be denied life itself; we would also be denied our weather. If, somehow, the earth's atmosphere continued to stay in place and continued to be purified and replenished even though it were denied the solar influence, that atmosphere would be uneventfully stable. Never heated, never tossed about by hurtling winds, there would be no storms, no changes in air pressure, no rain, no snow, not one of those electric days of September when the northwest wind gusts across Canada and spills over the northeastern United States bringing with it the essence of autumnal exhilaration.

Of all the weather changes I have experienced, the arrival of the northwest winds in September is the most memorable. There are winds—memorable and forgettable—like it around the world: the mistrals of southern France, the monsoons of India, the typhoons of the South Seas, the trade winds of the oceans, and the chinooks of Alaska, Washington, and Oregon. They, and every wind, are the products of the sun's heat—the radiation that is absorbed by every earth surface, whether it be snow, water, grassy field, dry sand, or dense forest. Pushing through the atmosphere, that heat becomes trapped by it, stays close to the earth where the air is dense, and, rising and falling in invisible columns, begins to mix the air currents that become wind.

Air warmed by the earth rises; the cooler air above it spills downward to fill the partial vacuum left by the rising column. Because the equator receives the most direct and most steady portion of the planet's sunlight and sun heat, it is the equatorial zone where every wind is born, although it may not grow into its full character until it has reached a part of the globe thousands of miles away. Were it not for the earth's rotation, all primary winds would flow upward from the equator while cooler air from the poles would rush north and south toward the globe's center. In the Northern Hemisphere, we would know nothing but

Sunsets can be read for temperature changes. "When the sunset is clear, a cool night draws near" gives a proverbial summary of the fact that without clouds in the sky, the earth's heat escapes more quickly into the air after the sun sets, and the night will be cool. If the sun sets like a ball of fire at which you can easily look, the temperature will probably rise with the warm, dry air to come.

Sunrises tell you about the weather that has just passed and so give indirect signs, best read along with sunsets. "The evening red (pink) and morning gray is the sign of a bright and cheery day; the evening gray and morning red, put on your hat, or you'll wet your head," is one couplet that puts together the sun signs. A gray morning means the rain-heavy air has passed to the east and the last evening's pink means dry air is on the way. A sunny day is coming up. Reversed, the dry air has passed and wet air is coming. Bright sunrises and sunsets together mean continued fair, sunny skies.

[*Continued on page 102*]

99

Sunrise on the Marshes by Martin Johnson Heade (1819–1904).

Red Hills and the Sun, Lake George by Georgia O'Keefe.

*With clean air and clear horizon
(over water, on deserts, or high on a
mountain), the rare sun sign of a
green ray seen just before sunrise or
after sunset speaks of bone-dry air.
Because the green flash is so seldom
seen, it has taken on legendary
meanings. The Scottish highlanders,
for example, say that whoever sees
the green ray gains the power to see
into the real feelings of the heart.*

*Sun halos and coronas speak silent
language to the alert sky-watcher.
Halos, white radiant rings around
the sun, tell of light bending through
high-up ice crystals at the top of a
water-filled air mass. In summer,
they tell of rain soon to come.
However, in winter, when the crystal
clouds drop lower in the cold air,
they may not so surely mean rain.*

*Coronas, small colored rings
around the sun that are red on the
outside and blue on the inside, tell a
double story. An expanding corona
means that the water in the air is
evaporating and weather is clearing.
A tightening corona means rain will
soon fall.*

cold winds from the Arctic, constantly sweeping across the continent. But because the earth turns, that primary air flow is diverted and becomes a series of lateral currents affected by ocean currents, mountain ranges, the position of glaciers and continents.

There are, of course, local variations on the major and primarily westerly and easterly winds of the two hemispheres. There is the sun-warmed air of the valley which moves up a mountain slope, cools when night falls, and comes back down the mountain. The same cyclical air flow, engineered by the sun, is what cools the coastal plain during the summer. As the sun warms the land near the sea or a large body of water, the air rises and cooler air glides in from over the water to take its place. Stop for a moment to consider the simplicity of that natural air conditioning the next time you enjoy a summer evening's breeze while you are vacationing at the beach.

These kinds of unequal heating not only create the overall pattern of air circulation, they are responsible for other kinds of weather systems which are known, at least by name—thanks to television—to millions of people who have no idea how they are created. When your friendly TV forecasters speak of a high-pressure system here, or a low-pressure system there, they are talking about whirling masses of air generated by variations in the overall patterns of air circulation. Highs develop in places where the air cools, compresses, and sinks. Lows are formed by the wavelike action of two highs of different temperatures, a complex meeting of air masses that builds what becomes "weather fronts"—another expression we hear nearly every day, but which is also seldom fully understood.

In the temperate zones, the weather changes of each season begin with the changes in the angle of the sun. Because the heat lost, or gained, as the angle becomes less or more direct is the primary weather influence, a day in Maine or Minnesota in July can be 100°F., and 0°F. in January. Those variations, in turn, affect the movement of the highs and

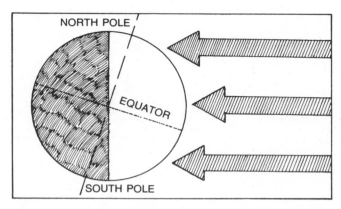

It's not because the sun is further away from us that the days get cold, but because in winter the sun's rays strike our part of the earth at a glancing angle. The angle means the light spreads out over a larger surface and each square inch gets less energy than when the sun is closer to straight on in summer. Thus, it is winter in the Southern Hemisphere, tilted away at an angle from the sun, at the same time that it is summer in the Northern Hemisphere, tilted to receive the sun's rays directly. At both poles, the rays ordinarily strike at an angle, the reason it never heats up there as at the equator.

Arrows show the solar rays striking earth when it is in summer position for the Northern Hemisphere.

In New Haven, Connecticut, a huge, old maple tree shows cosmic rhythms in its energy patterns. Yale scientists who wired up the tree for analysis over a 30-year period, found a 24-hour solar rhythm and a longer lunar cycle as well as variations produced by thunderstorms.

lows while they also determine if the water that is always a part of the atmosphere is converted to snow, rain, sleet, or hail when the clouds of any given weather system become too saturated with moisture to retain any more.

The water that transpires from the leaves of green plants is recycled and purified by the sun which draws it up as a vapor, leaving every impurity below. Our star works the same sort of distillation process on the waters of every river, lake, stream, and ocean, thus is born the term "relative humidity." If we could see the sun's hand in the conditions, we might be more content to abide by whatever weather comes our way. We would know as it does that we are witnessing a portion of the solar pageant that began with the unfolding of a new leaf on the maple tree in the village square.

We are, however, more inclined to view weather as a key to fashion, sport, or pasttime than we are given to thoughts about the role of solar energy on shaping the character of a summer shower or a winter flurry. We want to know if it will be cold enough to sustain the good skiing, or breezy enough for a sailing yacht. In this manner, the sun is reduced to a trifling presence, an energy source of interesting but frivolous dimensions, fine if it shines on a horse race, or covers the slope with powder snow, but undependable if you are counting on it to heat your home, power your machinery, or light an office.

Such misplaced evaluations are the result of our steady withdrawal from past intimate relationships with the natural world; we tend to forget the solar lessons learned by our agricultural ancestors of a century ago. They understood that it is the sun's warmth which sparks a dormant seed to germination; they could tell when the lengthening days in February meant it was time to expect their ewes to lamb. They could look at clouds—cirrus, cumulus, stratus, thunderheads, and tornado breeders—and know what the next day or the approaching night was likely to bring: rain, wind, snow, or hail. They understood the links

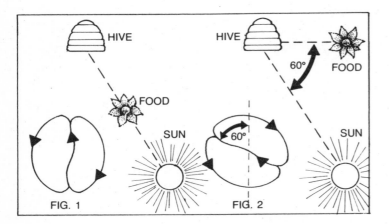

When a honeybee has found a good supply of nectar, she gets the message to other bees in the hive by a small dance containing a coded message about the relation of nectar, sun, and hive. Even on cloudy days, the bees can locate the sun. A vertical dance (figure 1) means the nectar is on a line with the hive and the sun. The angle by which the dance is shifted from the vertical tells the other bees the angle that the nectar is away from the hive-sun orientation (figure 2). The speed of the dance codes a message about how far to fly to the source.

between dew and frost, between raindrops, sleet pellets, and snowflakes. And when they looked skyward and saw their sun at the center of a crystalline halo of frozen ice crystals, they were warned that a storm was likely on the way.

For tens of thousands of other species the weather is everything. The breeding, molting, migratory, and feeding cycles of most birds, and even domestic fowl, is related directly to the hours of daylight in each day. Numerous experiments, using both regulated artificial light and measured amounts of sunlight, have shown that many birds depend entirely on the length of days (or the amount of light) to tell them when to begin breeding and thus insure the survival of the species. Similar "biological clocks"—which most often adhere to a 24-hour cycle, known to scientists as the "circadian rhythm"—can be found in most living organisms, including mice, monkeys, grizzly bears, sea worms, and the common fiddler crab, which darkens its body pigments during the day and lightens them at night.

In intertidal zones companions of the fiddler crab are particularly sensitive to the length of days and the span of darkness. Perhaps because they also observe the complex rhythms of the tides, caused by the interactions of the sun and moon and the gravitational pulls exerted by both, the shore dwellers have especially accurate biological clocks. Many marine algae—some of them among the most prolific of the photosynthesizers—also survive and flourish (and thus help support all life) by keeping time with light-sensitive biological clocks that are part of their living tissues. Just as bears and opossums hibernate, birds and whales migrate, and fiddler crabs and butterflys regulate their breeding cycles by the light of the sun and the phases of the moon, we are also creatures of celestial rhythms and solar and lunar presences. While there is still a great deal of mystery about how much influence solar rhythms have on a creature that lives in buildings with controlled climate and tinted-glass windows, it is generally acknowledged that a jet plane's

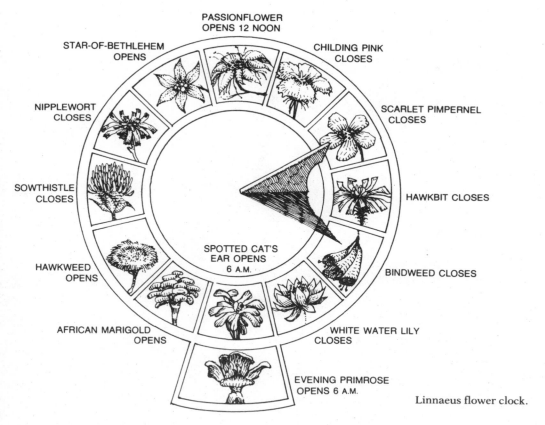

PASSIONFLOWER
OPENS 12 NOON

STAR-OF-BETHLEHEM
OPENS

CHILDING PINK
CLOSES

NIPPLEWORT
CLOSES

SCARLET PIMPERNEL
CLOSES

SOWTHISTLE
CLOSES

HAWKBIT CLOSES

HAWKWEED
OPENS

SPOTTED CAT'S
EAR OPENS
6 A.M.

BINDWEED CLOSES

AFRICAN MARIGOLD
OPENS

WHITE WATER LILY
CLOSES

EVENING PRIMROSE
OPENS 6 A.M.

Linnaeus flower clock.

The Swedish naturalist Linnaeus was the first to notice that flowers opened at different hours which could be timed as precisely as a clock. His formal garden, planted to form a clock face, let a stroller tell the time within a half hour by looking to see which flowers were open.

Later, scientists discovered that plants do light-detecting to regulate their flowering. Given the right day-length stimulus, the plant leaves manufacture a substance called florigen which switches the plant over from growth to flower-making.

Plants, however, are as much or more interested in the length of darkness than of daylight. Chrysanthemums, for example, can be held from flowering by shining a little light on them during the long autumn night—a break which tells the plants that the right time for flowering has not come yet.

ability to move "faster than the sun" across earth's lines of longitude is the primary trigger for the very real physical effects of what is commonly called "jet lag."

Seeds germinate in the warming spring soil, bears depart their dens, swallows build nests in the barn rafters, and whales cruise closer to the coast as the sun begins its return to the Northern Hemisphere. If we could learn more about precisely how the sun (and its sister/brother

[*Continued on page 108*]

The fallow field sparkled as the early sun touched the jewelled spirals of the great spider webs . . .
Mary Leister, *Wildings*

HAZE GOLD

Sun, you may send your haze
    gold
Filling the fall afternoon
With a glimmer of many gold
    feathers.
Leaves, you may linger in the
    fall sunset

Like late lingering butterflies
    before frost.
Treetops, you may sift the
    sunset cross-lights
Spreading a loose checkerwork
    of gold and shadow.
Winter comes soon–shall we
    save this, lay it by,
Keep all we can of these haze
    gold yellows?

Carl Sandburg, *Good Morning, America*

The Sower by Vincent van Gogh.

RESETTING A BIOLOGICAL CLOCK

MIDNIGHT: LIGHT TURNED ON

NOON: LIGHT TURNED OFF

CRAB OUTDOORS

SUNRISE

SUNSET

CRAB IN DARKENED ROOM

SUNRISE

SUNSET

Fiddler crabs change color on a sun-cycle schedule. Even when isolated from clues about light and tides, they transform themselves from a darker nighttime shade to a lighter daytime color, leading scientists to think that their rhythm is timed to an internal clock.

Insect clocks are set by the change from darkness to dawn, allowing tiny larvae or pupae to emerge into the outside world when humidity is at its highest near dawn or sunset. These rhythms also make insects more vulnerable to spraying at certain hours, about four o'clock in the afternoon for most flies and roaches, according to a government study.

Cockroaches, not the favorite friend of most house-dwellers, secrete a hormone in response to a light-to-darkness sequence. The hormone stirs them to activity two to four hours later, about the time most of us have left those dark rooms for bed.

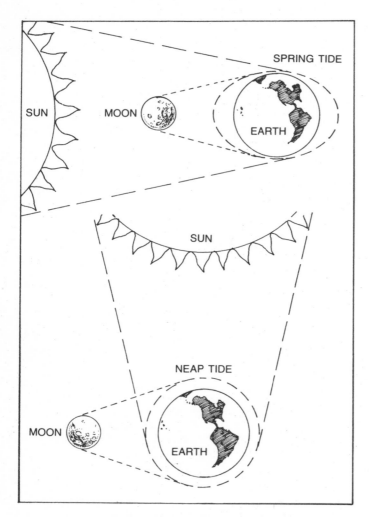

SPRING TIDE

SUN

MOON

EARTH

SUN

NEAP TIDE

MOON

EARTH

Angle of the sun in relation to moon and earth changes the ocean tides. The top diagram shows that when the sun and moon are in a line increased gravitational pull creates large spring tides. The bottom diagram shows that when the sun and moon are at a right angle the pull is less resulting in small neap tides.

moon) affects the behavior of animals and other living organisms, we would learn a great deal. If we could understand, truly understand, the ways weather is created, we would certainly develop more respect for the sun. If we could sense that there is more to be learned about the tides (and our predictions of these perpetual cycles are still relatively inaccurate), we might discover a good deal about ourselves and the creatures around us. For the sun's relationships to the natural world are still largely a mystery. We know, or say we know, a fair amount about *what* happens, but we are at a loss when it comes to explaining *why* it happens.

According to reports in the nation's leading magazines and newspapers, including editorials in the *New York Times*, the amount of carbon dioxide in the atmosphere is increasing at an ever-accelerating rate as the world burns more and more fossil fuels. The layer of carbon dioxide in the atmosphere allows the sun's rays to penetrate, but prevents the release of bounce-back radiation from the earth, just as a glass roof seals much of the sun's energy in a greenhouse. What this greenhouse effect will do to the planet's climate is the object of much speculation.

As public awareness is thus increasingly stimulated and informed, the public will come to realize that the sun is an alternative to fossil fuels and a resource which needs to be considered and protected as strenuously as any vanishing species or wild canyon. And, once that message becomes more firmly established, the sun can be seen for what it is and what it can be: a nonpolluting energy system which is already utilized by most of the earth's living organisms. Our species comes late to solar energy.

Once we have learned that there is no such thing as a free lunch, once we understand that burning fossil fuels and the sun's stored energy of centuries merely dims the star's present energy output, more of us will begin to comprehend the startlingly simple fact that we must care for the sun—and, in so doing, the earth also. One of the steps toward such an understanding is a more widely shared comprehension of the sun's

effect on every aspect of the vast, natural biomass around us—the biomass that sustains us on this planet.

Among these aspects of the biomass you will discover ozone—the rare gas (less than one-ten-thousandth of 1 percent) that makes up a portion of the stratosphere. This form of oxygen (O_3), with an extra atom, is scattered thinly in a high band about six miles thick. Reduced by ground-level atmospheric pressures, the ozone layer would be compressed to no more than a few tenths of a centimeter thick. A fragile and tenuous presence, ozone, nevertheless, plays a critical role in the survival of earth's organisms. Thin and scattered though they may be, the ozone molecules absorb much of the sun's ultraviolet radiation—radiation that could damage every living thing on a planet that has taken 15 billion years to adapt to the ultraviolet rays which do pierce the atmosphere. If the ozone layer is altered—and that is a process so little understood and yet so crucial that we could quite easily make the blunder—the entire global ecosystem could also be altered. Scientists have only theorized on the specifics of the trauma; generally, however, they agree that many more humans would contract skin cancer, and that the process of photosynthesis, the life-sustainer of the natural world, would be changed for the worse.

One of the questions which science has yet to answer (even though you may have thought otherwise) is the question of why the dinosaurs who walked the earth 65 million years ago failed to survive. They had, after all, evolved to a highly sophisticated and well-adapted group of animals. There is a growing number of researchers who now believe a diminishing of the ozone layer may have been the reason for the extinction of fully one-third of all the species that once existed. The researchers suggest that a supernovae, the explosion of a star in space, produced large amounts of nitrogen oxides, which destroyed the ozone shield.

We may have begun the journey to extinction in Alamogordo,

Ultraviolet energy from the sun, tearing electrons away from molecules in the upper atmosphere, creates electrified layers of air (called the "ionosphere") that make radio programs possible. The ionosphere acts as a reflector, bouncing radio signals back and forth between earth and sky.

Think of the ionosphere as a sieve whose mesh screens the waves sent up, bumping back the longest radio waves and allowing the shorter ones to disappear out into space. Changing solar radiation in the daily and yearly sun-cycles shrinks and expands the mesh.

Solar flares, sunspot cycles, and magnetic storms shake up the ionosphere as giant clouds of radiation shoot out from the sun to the earth's atmosphere. During that turbulence, distant radio signals may disappear completely through the changing size of the atmospheric mesh.

Along with helping our radios play, it almost seems as if the sun is playing with our radios.

All who have known the mountains have waited for the alpen glow, when the snow-clad peaks turn red for just a moment and glaciers and streamers of ice hang like colored ribbons from the heights. Long after the valleys are dark, those peaks continue to glow, and as the light recedes, they fade to purple and then to black with only the highest pinnacle flaming to the end. . . .

The Ross [sunset] light does strange and wonderful things to all who see it and are sensitive to its meaning. No two ever see it alike, but this much is true: somewhere within it is a power that transfigures everything, even those who watch.

Sigurd F. Olson's Wilderness Days

New Mexico, more than 30 years ago. Describing the first explosion of what was then called the atom bomb, Brigadier General Thomas Farrell wrote:

> The effects could well be called unprecedented, magnificent, beautiful, stupendous and terrifying. No man-made phenomenon of such tremendous power had ever occurred before. The lighting effects beggared description. The whole country was lighted by a searing light with the intensity many times that of the midday sun. It was golden, purple, violet, gray, and blue. It lighted every peak, crevasse and ridge of the nearby mountain range with a clarity and beauty that cannot be described but must be seen to be imagined. It was that beauty the great poets dream about but describe most poorly and inadequately. Thirty seconds after, the explosion came first, the air blast pressing hard against the people and things, to be followed almost immediately by the strong, sustained, awesome roar which warned of doomsday and made us feel that we puny things were blasphemous to dare tamper with the forces heretofore reserved to the Almighty. Words are inadequate tools for the job of acquainting those not present with the physical, mental and psychological effects. It had to be witnessed to be realized.

What the general, and the others who were impressed with that test effort at bringing the sun to earth, may not have realized is the effect of such explosions on the stratosphere. The nuclear weapons that have been exploded since have generated enough nitrogen oxides to deplete the

ozone layer by at least 4 percent, according to scientists who keep track of such matters. It is a wry irony of nature, is it not, that in our quest to take over the functions of the sun, we run the risk of destroying the fragile shield that now protects us from that star's ultraviolet rays.

The sun purifies our water, cleans and renews the air we breathe, sustains the life of the world's wild creatures, feeds the seas, builds rain clouds, stirs the crops to blossom, sets the winds on their course, defines the seasons, inspires our poets, nurtures the forests, restrains the polar ice caps, pulls at the tides, warms the robin's nest and triggers its construction, melts the snow, and creates, almost as an afterthought, the fuels that have advanced the Industrial Age.

There is no aspect of the natural world around us which is not a ward of the sun, dependent on the star for every vital detail woven into the tapestry of life that spans birth and death. And, it is that natural world which sustains us—the only creature in it who has suggested that he can build his own suns, that he can proceed even if, and when, he assigns a minor role to the sky's solar presence. It would be only prudent, would it not, to learn more about the sun's delicate arrangements with the natural world before we go further along the path of attempting to assume even greater dominance over that natural world. There are, surely, aspects of the natural sun we should know more about.

THE USES OF LIGHT

It warms my bones
*   say the stones*

I take it into me and grow
Say the trees
Leaves above
Roots below

A vast vague white
Draws me out of the night
Says the moth in his flight–. . .
Gary Snyder, *Turtle Island*

SECRET ATTRACTION

A rainy day: the hollyhock has
*   turned*
From east to west, as though the
*   sun still burned.*
Basho, *Chime of Windbells: A Year of Japanese*
Haiku in English Verse

The sun that warms you here shall shine on me.
                                                    SHAKESPEARE

4 *SUN PERSONAL*

As the architect of our weather, the sun does have an effect on our daily lives, whether we live in the city, suburb, or on a country lane. As the generator and transmitter of a spectrum of visible and invisible rays, the sun has an even more significant, less understood, and more hotly debated influence on our health, growth, psychic well-being, appetite, skin tone, and, some scientists will argue, our fertility and sexual vigor.

When the sun is bright, the humidity low, and a refreshing breeze blows from the northwest, the mood and behavior of people from Arkansas to Australia is noticeably improved and demonstrably congenial. Hot, sticky days spawned by a summer sun tend to shorten tempers and encourage violence. As city police have long known, and as their records indicate, there is always an increase in assaultive behavior when temperatures and humidity are high.

There are documented physiological studies which explain some of the reasons why the sun and the weather it orchestrates have such direct effects on our personal behavior; but there are also solar mysteries which have yet to be solved. The sun's brilliance, or obscurity, the sun's influence on high- and low-pressure systems, the sun's work at creating and dispelling the clouds overhead can determine your mood, your inclination, or disinclination, to work, the frequency and intensity of headaches, and the supply of oxygen to your brain. In warm weather, for example, your blood vessels expand to help the body shed its excess heat.

A Plains Indian mother holds her baby up to be blessed by the rising sun.

. . . under the glorious December light, as happens but once or twice in lives which ever after can consider themselves favored to the full, I found exactly what I had come seeking, what, despite the era and the world was offered me, truly to me alone, in that forsaken nature. . . . It seemed as if the morning were stabilized, the sun stopped for an incalculable moment. In this light and this silence, years of wrath and night melted slowly away. I listened to an almost forgotten sound within myself as if my heart, long stopped, were calmly beginning to beat again. . . . In the middle of winter I at last discovered that there was in me an invincible summer.

Albert Camus, "Return to Tipasa"

On cold days, they constrict to keep warmth in. As the diameter of your blood vessels changes, so too does your body chemistry, the supply of blood to your brain, and, in turn, your behavior.

However, available evidence based on studies of air-conditioned offices in New York and other cities, shows that attempts to circumvent the natural weather with artificial environments can have damaging effects. Workers in air-conditioned quarters tend to have more ailments, feel more uncomfortable, and have more headaches than those employed in naturally ventilated spaces. The same negative balance holds true for those who take artificial stimulants and drugs as innocent as aspirin. Because, like alcohol, aspirin dilates the blood vessels in the skin and quickens the loss of body heat, it may not be a wise move.

The sun influences the stability—or instability, in bad weather—of mental patients; its warmth can lull the driver of a car to drowsiness; or it can invigorate a university student who finds stimulation in the cool brightness of a clear October morning. And beyond these dimensions, according to Dr. Alfred Kinsey, author of the landmark sexual study, the sun can act as an aphrodisiac by stimulating adrenaline secretions which, in turn, stimulate metabolism so that more oxygen is delivered to the vital tissues of the brain, the heart, the kidneys, and the genitals. Fresh air and sunshine, the researchers agree, are two of the only true aphrodisiacs, the others being good food, adequate sleep, and regular exercise.

It is doubtful if Henry Thoreau had read any scientific papers or articles in the popular press of his day about the effects of the sun on the human condition. Nevertheless, his instincts as a man close to the natural world prompted him to recognize a link between his creative energies and his environment. Every day in preparation for his writing, he walked miles through the woods and along roads and footpaths wherever he happened to be. As scientists of a century later are beginning to discover, such direct, outdoor exposure to the sun can have physical

Because many people have experienced vision improvement from sunning, Lisette Scholl in her book, Visionetics, *describes a safe exercise for beginners.*

With the eyes closed *and at a time of day when the sun's intensity is low (early morning or late afternoon), sit or stand in direct sunlight. Slowly move your head as if you were circling the sun with your nose, making the movement about five times in each direction—clockwise and then counterclockwise. Breathing deeply, yawn and relax while you do this, giving in to the sensation of the sun's warmth. "Let the sun's power totally encompass you as you absorb the primal energy of our solar system."*

What is called a physiological clock may be . . . the organization that knits together the human race, and binds it firmly to the earth and vast universe beyond.
David Sobel, *Ways of Health*

and physiological effects which are only now beginning to be discussed, much less fully understood.

Like most typical citizens of the automobile age, I do not walk hours before I get to work. Nevertheless, like most of us, I have learned that sunlight and its concurrent "good" weather does indeed have an

Graditude to the Sun: blinding
pulsing light through
trunks of trees, through
mists, warming caves
where
bears and snakes sleep—
he who wakes us—
in our minds so be it

Gary Snyder, "Prayer for the Great Family"

effect on my work habits and my general behavior. The reflection of the sun on blue water, the patterns of sunlight and shade on a tree-shaded country road, the rolling of the rising sun's rays across a rippling field of wheat, the gleam of the sun on a snowdrift, and the brilliance of the sun in the western sky as I leave my office for the journey home—these are the solar aspects of a typical day and they continually inspire positive feelings. When the sun is hidden by storm clouds, or during the darkness it leaves in its wake as it shines on the other side of the world, I feel a different rhythm, am prompted to different, more subdued activities. I am influenced as much by what I cannot see as by what I can, and I welcome the sun when it returns to my home skies once again.

These are not peculiar or exceptional responses; they are, I am sure, shared by hundreds of millions of men and women in the temperate zones of the world. The sun has been our star for too long to be ignored, to be taken for granted, to pass unnoticed, even in a world where technology promises, but often fails to deliver, benefits almost as universal and munificent as those of the sun itself. What we know, what we sense beyond any doubt, is that the sun makes us feel better, makes us view our existence here with more optimism, with more confidence that this day will amount to something, that this day is worth living.

There is, I'm sure, a general nodding of heads in agreement. Like me, you have spent many moments in your life when you not only acknowledged the sun's presence, but were cheered, awed, and gratified by it. Also, like me, you may not have known that there may well be valid scientific reasons for our positive feelings about the sun. As yet, very little exploratory work has been done on the surprising ramifications of sunlight's full effects on the human mechanism, but enough has been discovered to make it clear that our instinctive yearning to be in the sun may relate directly to an issue as fundamental as survival.

In addition to the seven colors of the rainbow—those color components of sunlight which you can see in a rainbow's arch—and the

THROUGH HIS EXTENSIVE EXPERIMENTS WITH LIGHT,
SIR ISAAC NEWTON (1642-1727) BECAME ONE OF THE
FOUNDERS OF MODERN OPTICAL SCIENCE

© BAUSCH & LOMB OPTICAL CO. ROCHESTER N.Y.—AMERICA'S LEADING OPTICAL INSTITUTION

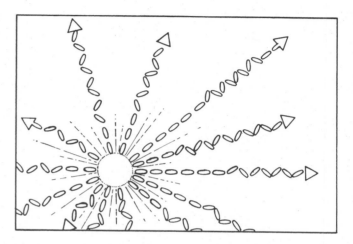

Today we picture light as due to the motion of
tiny bundles called "photons."

Sir Isaac Newton (1642–1727) dispersing white
light into colors using a triangular glass prism.

Despite its familiarity as it plays daily over our earth, the nature of light is still embedded in mystery. To explain what it really is, a long history of scientists have formed particle and wave theories at the heart of which still pulsates an unknown essence.

The ancient Greeks observed the straight shadows of houses and trees on the ground, outlined as if light traveled in straight paths through the air. This together with the straight, angular reflection of light from mirrors led them to speculate that light was a stream of particles flowing out from the sun. For centuries philosophers used the Greek particle theory to explain the actions of sunlight and shadow, reflections, and mirages.

Already I began to love the sun;
A boy I loved the sun,
Not as I since have loved him,
*   as a pledge*
And surety of our earthly life,
But for this cause, that I had
*   seen him lay*
His beauty on the morning hills.

William Wordsworth

eight different wavelengths each represents, the sun also sends us some invisible rays, some noncolors which are turning out to be just as important to this planet as the white light of the sun's spectrum which conquers darkness, allows us our colors, and defines beauty. The world around us exists visibly because rainbows exist; perhaps that is why they are such a universal symbol of hope and faith in the future.

The sun's messengers we cannot see are radiations beyond the frequencies of the colors in the rainbow. They have wavelengths just below the violet at one side of the spectrum, and just above the red at the other. There are gamma rays—with the shortest of wavelengths—and low-intensity x-rays. These are but a small fraction, however, of the sun's invisible messengers. The most significant two are the ultraviolet rays (just a bit shorter than the violet on one end of the spectrum) and the infrared rays (just a bit longer than the visible red on the other end).

The infrared radiations make up the majority of the solar package—about 60 percent. Visible color, the white light that we can see in its total and prismatic versions, is about 37 percent, and the so-called ultraviolet (beyond the violet we can see) rays that get through the ozone layer account for the balance, about 3 percent. We have a great deal left to discover about the precise effect these solar messengers have on the life of the planet and its creatures—including each of us.

So little, in fact, in the way of absolute understandings of the nature and effects of ultraviolet rays has been established that it should come as a surprise to anyone with a sense of logic that we have tampered as much as we have with our atmosphere—the only practical regulator (and a fragile one at that) of the sun's ultraviolet emissions. Yet we are adding pollutants to that atmosphere at a rate that is certainly able to alter the very structure of our only shield and filter.

Which is not intended to imply that all ultraviolet rays can be assumed to be harmful to your health. There are reputable scientists who will argue that such rays should be treated with much circumspection

Odin at the foot of Asgard, the "rainbow bridge" leading to Valhalla, abode of the gods.

Cultures across the world have woven myths around the magical rainbow. In many, the bow is a link between heaven and earth. Japanese gods walked upon the "floating bridge of heaven" as they reached down to create the islands. Norsemen called the bow the "bridge to heaven" and North American Indians called it the "pathway of souls." For the Greeks, the arc of the perfect circle was the pathway for the rainbow goddess Iris, swift messenger of the gods to men.

The biblical story of the Great Flood which destroyed most of mankind captures the rainbow's promise of harmony after the struggle of a great storm. In it, God speaks to Noah: "I do set my bow in the cloud, and it shall be a token of a covenant between me and the earth . . . and the waters shall no more become a flood to destroy all flesh." Medieval alchemists, for whom the transformation of lead into gold was a symbol of spiritual growth, used the rainbow as a sign that the struggle between the physical elements was over and harmony reigned.

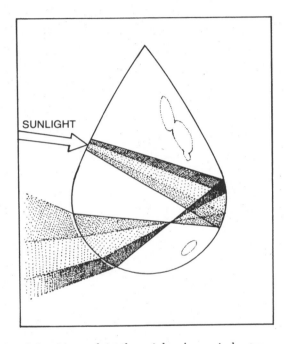

Scientists explain the rainbow's magic by tracing the path of light through a raindrop. As a ray of light strikes the drop of water, it doesn't go through and out the other side, but is reflected from the back of the drop. The light ray is bent twice, once as it enters the drop, and a second time as it goes out after bouncing off the back. Just as in a prism, the colors of white light separate into bands of the spectrum: violet, blue, green, yellow, orange, red. Because each color has a different wavelength, the light is separated into bands from violet to red.

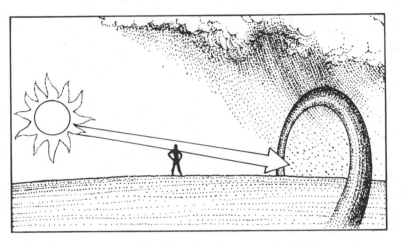

A rainbow becomes visible when sunlight is reflected through falling mist or rain.

because, on the whole, ultraviolet emissions are more trouble than they are worth. There are, however, equally forceful and documented arguments from equally reputable scientists that UV rays (as they are called) can have benefits beyond any of medicine's current horizons.

To begin with what has been acknowledged over the longest time span as one of the most important functions of the sun's invisible rays, one should read the history of rickets and the subsequent determination that sunlight and vitamin D play a key role in the healthy growth and development of most of the planet's vertebrates—including human beings. Rickets has been called the first disease of the Industrial Age; it reached epidemic proportions in earth nineteenth-century England at the start of the Industrial Age when the burning of large amounts of soft coal created so much soot in the atmosphere that it obscured sunlight. In suburbs of London, Liverpool, and other northern European industrial cities and towns, rickets affected thousands of children. Without sunlight and vitamin D, their bones developed improperly, their legs bowed, their muscles ached, and eventually their livers and spleens began to degenerate.

The same ailment bothered young animals in the London zoo. Chimpanzees, chipmunks, lions, tigers, bears, deer, ostriches, and even lizards exhibited the same symptoms as many of the children who peered at them through the bars of their cages. When cures for this affliction were found, it was discovered that sunlight was the most dependable. If that could not be assured, exposure to artificially generated ultraviolet rays also did the job, as did the consumption of cod-liver oil, known to be rich in vitamin D. In the twentieth century, vitamin D has been synthesized through a photochemical process involving ultraviolet rays that simulate the sun's.

Among the mysteries encountered during the discovery of the causes and cures of rickets was the high vitamin-D content in fish oils. As creatures of the ocean depths, fish are not thought to be exposed to enough sunlight to generate the vitamin concentrate, nor do scientists believe that they can absorb it from their diet of other, smaller fish or marine algae. The most-often-stated theory suggests that fish have developed an internal metabolic process that somehow extracts, stores, and perhaps fortifies, the vitamin D that other creatures get from the sun. What happens beneath the sea, however, is still largely a matter of conjecture. It wasn't until mid-1979 that exploring botanists in the Arctic discovered plants growing on the ocean bottom beneath a permanent ice cap some 18 feet thick. Until the plants were revealed in actuality by underwater cameras, most scientists would have told you quite firmly that there was no way such vegetation could survive at such depths and in such relative darkness.

Some hints as to how the entire evolutionary process has been

Suddenly a green and rose rainbow shafted right down into Starvation Ridge not three hundred yards away from my door, like a bolt, like a pillar: it came among steaming clouds and orange sun turmoiling.

What is a rainbow, Lord?
A hoop
For the lowly.

It hooped right into Lightning Creek, rain and snow fell simultaneous, the lake was milkwhite and a mile below, it was just too crazy. I went outside and suddenly my shadow was ringed by the rainbow as I walked on the hilltop, a lovely-haloed mystery making me want to pray.
Jack Kerouac, *The Dharma Bums*

Shine! Shine! Shine!
Pour down your warmth, great
*   sun!*
While we bask, we two together
Walt Whitman, "Out of the Cradle Endlessly Rocking"

"*The morning needs to be sung to.*" *Look east and welcome the sun with whatever song arises in your heart as you look. "If the sky turns a color sky never was before, just watch it. That's part of the magic. That's the way to start a day.*"
Byrd Baylor, *The Way to Start a Day*

shaped by the sun were found in the rickets outbreaks of industrialism's early days. In addition to filling the English skies with soot, the first of the mass-production manufacturers imported cheap labor from India and Pakistan. These dark-skinned children suffered most severely from rickets because the natural sunscreening mechanism in their skin cells—a dark pigment known as melanin—had evolved over centuries in their warmer and brighter latitudes to allow roughly the same amounts of vitamin D to build up in prolonged sunshine as English children could get from their shorter summers. In their heedless transport of the Asian children from one climate to another, the British colonialists failed to take the sun into account—and the results were tragic.

It is likely all of us will make more such misjudgments in the near future; the relationship between us and the sun is not clearly enough defined to prevent it. Current theory holds that ultraviolet rays are a primary catalyst in the process of cell regeneration—the basis of life itself. It is also known that the same UV-B messengers from the sun have an effect on the recently discovered DNA strands, the genetic material of the chromosome that is each organism's hereditary encyclopedia. We are who we are, geneticists tell us, because our DNA patterns decide it. Yet it is also known that those same DNA macromolecules are most sensitive to ultraviolet (UV-B) emissions. It would follow, then, would it not, that the sun is indeed our parent, both mother and father, male and female, determining the sex and characteristics of its children. The ancient civilizations of tens of centuries ago proclaimed as much, yet they left no records of DNA and UV-B discoveries. They left only carvings in stone of the sun's rays falling on this earth and its creatures. But in their awe of the sun, they did little to displease it. They were wiser than many of us may have assumed. There were no epidemics of rickets in ancient Egypt, nor was there any man-made destruction of the ozone layer, or any man-made addition of vast amounts of carbon dioxide to the atmosphere.

We are, as we have always, altering our environment, changing

our relationship with the sun. The difference between what we are doing today and what we did earlier in our existence is the current assumption that we know what we are doing.

Surely the industrialists of London and Liverpool never imagined that the burning of coal to energize their production lines would inflict twisted bones and damaged livers on their own children. Would the sun have been obscured by soot, if we, like the ancients, still looked to the sun as the parent of us all?

A hypothetical question, to be sure, and eminently unanswerable. Nevertheless, the image of the sun as parent is worth holding. Like most parents, it can support as well as discipline. There are mostly benefits, not damages, to be gained from our further re-acquaintance with the sun and our development of closer, one-on-one relationships with it. We are most likely to receive those benefits if we open our arms, and our minds, to the sun, instead of assuming that we can pollute the skies without altering the sun's subtle relationships to this planet and our general well-being.

John N. Ott, author of a sun book titled *Health and Light*, became acquainted with the sun as a photographer who specialized in time-lapse motion pictures of plants. As he recorded the opening and closing of flowers like the morning glory with high-speed cameras that "collapse" time, he became more and more knowledgeable about how the sun affected his subjects. He learned that it is the sun that sets the rhythms of plant life, that decides when the petals open and close, when seeds are fertilized, dropped, and then germinate to complete the cycle. As he came to understand how different lenses admitted different sorts of light into his cameras, and how plants moved to sunless environments reacted to artificial lights, Ott began to wonder about the ways in which sun could affect other living organisms, especially the one he was most interested in: his own.

Because his work as a cinematographer was his primary source

[*Continued on page 128*]

Twentieth-century theory describes light as patterns of energy particles called photons. Billions of infinitesimal photons, which nobody has ever seen, are assumed to form the electromagnetic waves we call light. Atomic reactions in the sun cause solar atoms to emit energy photons. The more energetic photons form blue light, the less energetic make red light, with the other colors in between. The assumption about photons explains the actions of photovoltaic cells in which light creates a flow of electricity by striking and dislodging electrons from the metal in the cell.

Yet even in twentieth-century theory, behind the names and beyond our vision, lies the still-mysterious essence of light, the gift of our sun.

Biologist C. R. Woese of the University of Illinois hypothesizes that living cells did not emerge from some primordial ooze, the "soup of the sea," but instead were created by the interaction of the sun's ultraviolet rays and myriads of tiny, salty droplets of water in the air.

"We have to take seriously," says Woese, "the idea that the earth was still so warm that there were no oceans. Indeed, one has to discard completely the notion that the oceans are the place where life began."

Woese bases his thesis on laboratory research done with ultraviolet light and gas mixtures. When exposed to the invisible portion of the sun's rays, the gases are changed by a chemical reaction and produce amino acids, the important building blocks of protein. The chemical event is a slim thread to the origins of life, but enough has been learned to allow respectable scientists to debate it.

The Ancient of Days by William Blake.

126

Huichol Indian yarn painting, *Tatei-Urianaka,
Goddess of the Waters.*

127

In an ancient Hopi ritual, people greet the morning sun facing east as it begins to rise above the horizon. The first deep yellow to appear is thought of as pollen from the sun which is symbolically scraped off the horizon with a gesture of the hand and put in the mouth as food. As the sun rises further, four deep breathes cleanse heart and insides. Then four times a hand movement over the body symbolizes clothing oneself in the power of the sun. And finally, as the sun is fully risen, the Hopi observing the ritual reminds himself to keep his own face as benign and powerful as the sun's.

I think continually of those who
    were truly great.
Who, from the womb,
    remembered the soul's
    history
Through corridors of light where
    the hours are suns,
Endless and singing . . .

Born of the sun, they travelled a
    short while toward the sun
And left the vivid air signed with
    their honour.

Stephen Spender, "I Think Continually of Those"

While the sun and the moon may have their halos and their coronae, we human beings can have a glory around our own heads–or, more accurately, around the heads of our shadows. If you stand on a hillside when the sun is low in the sky and clouds are hanging close behind you, you can see your own shadow outlined against the mist. The head of the shadow is surrounded by a glory in vivid spectral colors. If you are standing with a friend, you can see his shadow, too, but the glory is only around your head. . . . Another example of the glory is seen when a plane, flying toward the sun, emerges from a cloud. The passenger, looking back, can see the shadow of the plane "caught in a noose of light."

Louise B. Young, *Earth's Aura*

of income, and because the nature of that work required him to stand relatively still for long, long hours, Ott was most concerned about his developing arthritis; it was not only discomforting, but it began to hinder the primary creative work of his life. He took what he had learned from that work and what he suspected about sunlight and tried a personal experiment. He stayed away from artificial light as much as possible; he avoided wearing glasses or sunglasses, or even spending much time behind the windshield of his car, because all standard glass screens out ultraviolet rays, and he spent as much time in the sun as his work would allow, even in cold and cloudy weather.

The results, as he reports them, were significant. "The effect on my arthritis [he writes] was as beneficial as an injection of one of the glandular extracts right into the hip joint, but without the intervening day or two of discomfort. There was no doubt about it. My arthritis was definitely much better, and I was satisfied it was not imagination or wishful thinking. Furthermore, after several days of not wearing glasses at all, my eyes were no longer so extra sensitive to the bright sunlight, even on the beach. Before the week was up, I played several rounds of golf on a short, nine-hole course and went walking on the beach without my cane. I felt like a new person."

There has been much spirited debate in the medical fraternity about this, and the reason for debate is that there is so much room for it. No one has stated absolutely, as a medical fact, that the sun affects this organ in that specific way. Instead, most medical specialists will admit that there is little known as fact, and a great deal of work still to be done. Some of that work has begun—although not on any major scale—and as it continues, it has become clear to researchers that they need to learn more about the pineal gland, the pituitary gland, the endocrine system, and the hypothalmus. Of these, the pineal gland—a small organ positioned at the center of the brain—is getting the most early attention. It is known that it is affected by light that enters the eye. Just how is not

known, but Ott believes that the sun's ultraviolet rays are the key to pineal stimulation—which is the reason why he would not allow glass to come between him and the sun. Just how the pineal gland functions, on the other hand, is even less clearly defined. It is, however, in concert with each of the other glands and systems, a part of the human machine, and functions in a wondrous concert with each of them.

One scientist, Dr. Richard J. Wurtman of the neuroendocrine laboratory at the Massachusetts Institute of Technology, basically agrees with Ott and says that sunlight affects us through photoreceptors in our eyes just as much, if not more, as it affects the cells of our skin. Sunlight striking the retina, he says, stimulates the optic nerve, which, it seems, sends impulses to the hypothalmus—a part of the brain with a great influence on the emotions. From there, stimulation travels through neurochemical channels to the pituitary and pineal glands, which, in turn, release the hormones that help control body chemistry.

Ott also believes that sunlight has a direct affect on cancer. Perhaps the most well-documented and startling of his case studies is one concerning a large group of schoolchildren. As he describes it: "According to the U. S. Health Service, a school in Niles, Illinois, had the highest rate of leukemia of any school in the country. In fact, it was five times the national average. I made a point of visiting the school and talking with the superintendent, the head maintenance man, and also some of the teachers who had been at the school since it was built. I learned that all of the children who developed leukemia had been located in two particular classrooms, and that the teachers in these classrooms customarily kept the curtains drawn at all times across the windows because of the intense glare from the extensive use of glass in the new, modern building. On examining the curtains I found that they were not completely opaque, but more of a translucent type of material that allowed some of the outdoor light to penetrate and which gave them a greenish appearance. With the curtains constantly closed, it was necessary to keep the artificial lights on

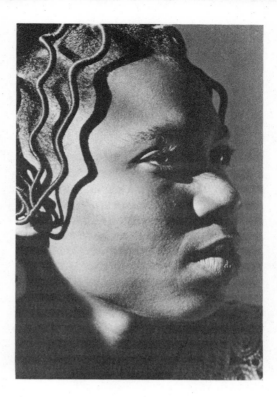

Light is sweet; at sight of the sun, the eyes are glad.
Ecclesiastes 11:7

The eleventh-century Arabian physician Avicenna diagnosed disease based on the color of eyes, skin, and excrement, and developed a description of temperament based on colors. The red-haired person was volatile, for example, while the brown-haired were cold in personality. Blue light slowed the movement of blood, he thought, while red light stimulated it.

Often we're aware of our body clocks only when they're thrown off rhythm by sudden changes, but in fact they're working all the time. Babies in the womb tend to kick at a regular time each day. Once born, we have over a hundred functions and elements that show a daily oscillation. The volume of urine excreted, our reaction to drugs, and our speed in solving mathematical problems all vary depending on time in the daily sun-cycle.

People react differently to disease or surgery at different times of the day. And just as our body clock may wake us up at 6:30 A.M. each morning, so some "gut" emotional experiences like fear may become linked to a certain time in the day–night rhythm.

There are lots of practical impacts of those rhythms. People who want to lose weight, for example, should probably eat their heaviest meal in the morning because the rhythms in protein formation make that the most efficient time of the day for using them.

Awareness of our 24-hour body

[Continued on page 131]

in these two classrooms, and I learned from the head maintenance man that the original tubes installed were 'warm-white' fluorescent, which are very strong in the orange-pink part of the spectrum.

"After several years of this regular procedure of keeping the curtains closed and the lights on, the classroom teachers of these two particular rooms left the school; their replacements preferred to leave the curtains open and the lights off unless they were needed. I also learned that about this time there was a general replacement of the warm-white fluorescent tubes and that the new tubes were cool-white, which are not as strong in the orange-pink part of the spectrum, and which were not lit continuously, but only as needed. As of the time of my visit in 1964, there had been no further leukemia cases reported for several years. No explanation for the previous, unusually high rate had ever been found, but the problem no longer existed. The situation had returned to normal."

One can understand why Ott would be criticized by the medical fraternity for using this example to demonstrate that artificial lighting of a particular sort can aggravate the incidence of cancer. Or, put another way: the absence of sunlight and natural ultraviolet emissions greatly increases the chance that a young person will contract leukemia. Ott makes no such flat claims, but he continues to argue that more exploratory work needs to be done, and he pursues his convictions about light and health with much persistence.

Most of us who work in places that are illuminated by fluorescent lights (90 percent of the work force, according to one statistic) are more affected by Ott's work than we might imagine. After 10 years of trying, and after turndowns from two leading manufacturers of fluorescent bulbs, the "full-spectrum" bulb that Ott had developed with sunlight as his model, was marketed. Since then, testimonials from manufacturers and industrial plants which have installed the bulbs support what Ott has been saying for the past 20 years: the light we live in makes a greater difference to our health and well-being than most of us suspect. Some

"This is how I like my energy. Straight!"

remarkable testimonials have come from the "full-spectrum" employees and their managers. There has been less absenteeism, lower accident rates, and marked increases in production.

It is typically American and quite inconsistent with the free-enterprise system that indoor lighting inspired by the sun should be accepted because it results in better worker performance on the production line. Perhaps that is the way the entire effort to affirm the practicality of solar energy should begin. If it can be demonstrated that in addition to all else that it does, the sun can also increase profits, then there will no longer be any debate about the nation's energy horizons.

Such a strategy will not be used on a large scale in the near future, however. In light of what is already known about the sun's benefits for the individual, one can only lament the existing energy systems—nuclear and fossil fuels—which prevent us from turning more quickly to the sun, and one can only hope that more and more people like John Ott will continue to publish and report on their increased under-

rhythms is especially important if we need to make abrupt changes in our life such as flying across many time zones or changing to night work. Several days are needed to resynchronize biological rhythms after flying through three or four time zones. Air traffic controllers, nurses, and factory workers who frequently rotate shifts may suffer from the body stress of changing demands on their system.

. . . of all the familiar phenomena of Earth, light now seems . . . to be more nearly absolute than anything else. Not only does it travel mysteriously in the vacuous dark, its velocity the almost unbelievable constant that has been made a cornerstone of modern physics, but its immateriality is probably the most measurable connecting link between the physical and mental worlds.

Guy Murchie, *Music of the Spheres*

To sit in the sun. This is still one of the greatest experiences of life . . . And it is free. . . . At the chosen moment I lay down, curled up, and closed my eyes while the sun shone on my face. Often a strong chilly wind blew, but it didn't come near me, I received only the sun. Then I entered my own special, simple paradise. . . . Let the world outside carry on, I would say, let them dash hither and thither, let them kill one another wholesale, let them go to hell, I'm wrapped in the embrace of Nature and filled with peace and love! And like any dog, like any savage, I lay there enjoying myself, harming no man, selling nothing, competing not at all, thinking no evil, smiled on by the sun, bent over by the trees, and softly folded in the arms of the earth.

John Stewart Collis, *The Worm Forgives the Plough*

standings of the often mysterious relationships between us and our star.

It would be ironic indeed—and a feather in Ott's often maligned cap—if the ultraviolet messengers of the sun are ever proven to be the triggers of all life on the planet. The hypothesis, if it gains general public acceptance, would do much to change the Dr. Jekyll–Mr. Hyde image that sunlight has acquired in these days of dark glass, suntan lotions, sunbathing clinics, and the prestige of the perfect tan.

"Too much sunbathing can be harmful," is the conventional wisdom, true enough in its basic message: too much of anything almost always bring some form of grief to the heedless consumer. But by constantly stressing this message about the sun, writers of the popular press may have helped to obscure other information which can persuade all of us that our star is more than just a natural cosmetic that helps white Americans look sleek, healthy, and rich through winter and summer.

It is true that the interreaction of certain ultraviolet rays on the skin's melanoma cells does alter pigment—in some cases so efficiently that the blisters and peeling of sunburn are the result. It is also true, according to accepted medical theory, that overexposure to the sun can cause some types of skin cancer—a theory many argue can be proven by the higher skin cancer incidence among farmers, loggers, fishermen, and other occupational groups whose work requires them to be outdoors most of the time. A question still to be explored, however, relates to the factors that may contribute to what appears to be a higher incidence of skin cancer since the onset of the Industrial Age. Do the pollutants industry adds to the atmosphere, one might ask, alter the effects of the sun's rays? And is it possible that if they do, the "altered" sunlight is causing skin cancer? Researchers concerned with the issue might wish for a time machine that could take them back to a preindustrial time. Then they could collect sunlight and skin statistics unmarred by the effects of air pollution. The results might be surprising.

*Before turning in, we step outside
for a quick look at the night sky, but
we rush back for our parkas and
mitts. The northern lights are out in
full canopy and have taken over the
heavens. A sky full of dancing
motion as the lights race across the
blackness, erupting in sudden streaks
from behind the mountains, creating
a spectrum of ephemeral jewel
colors—intense emerald greens, deep
reds, golds. Shimmering, incessantly
moving, changing, vanishing,
reappearing, they congregate finally
in an iridescent canopy just over our
heads.*

*Our initial cries of "Look
there!" . . . Did you see?" . . .
Look!" soon subside, and we stand
silent, enveloped in a wild and
magical event of light and color.*
Billie Wright, *Four Seasons North*

The aurora borealis silhouettes an Alaskan
forest. Often called "northern lights," this
stream of charged particles from the sun in-
teracting with earth's atmosphere is one of winter's shimmering shows in the Far North.

133

Color is not inert. It is the result of a constant bombardment of the earth by the spectrum of light from which each bird and flower absorbs the wavelengths suited to their nature and bounces off the rest. What is bounced off, we see as color.

A red geranium has absorbed the short blue waves and turns back the red; the bluebird has taken in the longer red waves and flings back the blue. Highly polished surfaces reflect almost all light like a mirror; a piece of charcoal absorbs all rays and appears black.

In each cubic millimeter of air, there is sufficient room for a million bacteria to grow. Without sunlight, the natural enemy of bacteria, we should long since have been stifled by the fast-multiplying microorganisms. Each day when the first ray of sunlight strikes the earth, its ultraviolet radiation paralyzes the bacteria, robbing them of their ability to propagate, and clearing space for us on earth.

Man and Dog in Front of the Sun by Joan Miró.

Meanwhile, the topic of sunburn is the primary one for thousands of solar scientists. On a much-less-publicized level, a few biologists and physicians are studying other aspects of sunlight and its relationship to human health.

A Boston physician and clinical researcher at the Massachusetts General Hospital, Dr. Robert Neer, is working on a theory that will help find the reasons why older people's bones become brittle and misshapen—causing, among other ailments, the old-age stoop. Dr. Neer believes that like the children of nineteenth-century London and Liver-

pool, the older people of our cities may not be getting enough vitamin D, or may not be able to absorb it as readily as younger people. He is treating a number of test patients at a VA home with extra portions of ultraviolet light and sunlight as well. He has found that the regimen does produce more vitamin D, but, as he says, "We need much more information."

Those words are echoed by every individual who has worked or who is currently working on trying to determine the precise ways sunlight

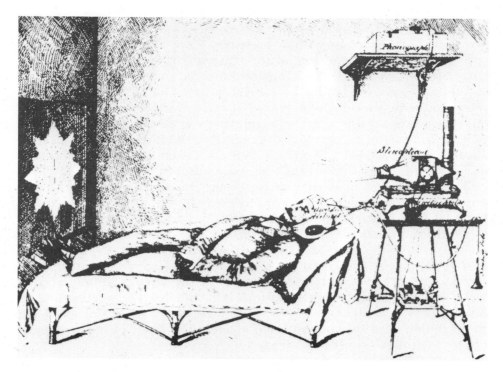

Victorian color therapists treated illness with musical vibrations and color slides.

Imagine that you suddenly gave up eating all fruits, vegetables, grains, nuts, and meats, and began eating pasta, candy, and sugary cereals only. All these groupings are "food," but the nutrients within each are substantially different. Where they are the same—there is some protein, for example in candy, and there is starch in some vegetables—they are of entirely different proportions. Eating pasta, candy and cereal will keep you alive, but eventually it will affect your health. And so it is with alterations in light-diet from the "natural" mix of spectral ingredients to the artificial mix.

Jerry Mander, "Four Arguments for the Elimination of Television"

In the nineteenth century, color therapists developed a theory of the psychological as well as healing effects of color. Blue is restful while yellow stimulates activity and white gives a sense of equilibrium, they said. Red light was held to be valuable to combat paralysis and other chronic conditions while blue and violet soothed nervous conditions.

Indian midwives who placed jaundiced infants in the sunlight to cure them long predated the scientific discovery that sunlight was preferable to blood transfusion for this condition.

For the ancients, color provided a way to restore the healing balance of unity seen as the cure for disease. Red cloth, for example, was used to stop bleeding. The power of the sun was transferred to yellow-colored objects like gold or turnips for the cure of jaundice. Black snails were rubbed on warts in England.

Contemporary sciences show that color does affect our muscular tension, heart rate, and other bodily functions as well as arousing emotional reactions. Respiratory movement increases during exposure to red light and decreases with exposure to blue, found Robert Gerard, a California psychologist. Bright colors draw interest outward while softer, deeper colors let us relax and concentrate within.

affects the planet's creatures, especially humans. Russian studies have shown that children whose classroom lighting was supplemented with low-level ultraviolet were less hyperactive, and thus better students. And tufted puffins at the Bronx Zoo laid their first fertile eggs in captivity after the lighting in their zoo homes was changed to full-spectrum bulbs that gave the birds the entire range of rays they would normally get from the sun. The United States Navy is using the same lights in submarines and finds that its underwater sailors are measurably more healthy, both physically and emotionally.

We thought because we could unravel the sun's nuclear secrets at Los Alamos that we had learned all there was to learn from our star. We are just now discovering that the sun is more than a nuclear reactor for physicists to emulate in miniature. It is more than a tanning machine for practitioners of conspicuous consumption, a cosmetic for the beautiful people. It is, we are beginning to discover, a bestower of health, a cure for disease, and a monitor of the psyche that can indeed make us "feel good" about life, prompt cerebral activity, improve our demeanor, stabilize our emotions. Indeed, as we may discover one day soon, the sun may be able to ease our aggressions so effectively that a solar civilization will become one of enduring peace rather than recurring war.

We do not have to look to some nebulous time frame for the research that can support such notions. We already know, for example, that we are influenced psychologically and physically by the sun's 24-hour rhythms, although we are not absolutely certain about the whys or hows. Hospital data collectors have learned that deaths and symptoms of illness do not occur with regular, around-the-clock rhythms, but instead cluster at certain hours on the solar clock. For the same, still unanalyzed, reasons, more pregnant women go into labor during the night or early-morning hours. And all of us, as any jet aircraft traveler will tell you, have come to understand that our internal clocks—the ones set by the sun in our home latitudes—must be respected.

The Petrified Forest by Max Ernst.

We do, indeed, live by the sun. Those few peoples on the earth who must live long portions of their lives without it—like the Eskimos at the rim of the Arctic—suffer for it. Observations by Dr. Joseph Bolen of the University of Wisconsin among the Eskimos living in Wainwright, Alaska, cited a high incidence of emotional illness and a malady the Arctic people call "winter madness"—a hysteria that can send them on compulsive and often fatal journeys across the endless expanses of ice.

As I sit watching the gold-touched clouds in an evening sunset over the lake, watching as the gold deepens to orange and then pinks, finally purples and deep reds as the blazing globe itself sinks beneath the horizon, I become aware of the world of colors in everything touched by the sun's light. The line of gold rippling over the waves lights my heart. When I murmur in my sounded language a thanks to the sun who has spoken with the language of color to me, the beginnings of a relationship are forged. As I slowly become aware of its role not only in giving me color, but in shaping all life on earth, I become a being of the universe. Picturing myself as one point in a complex network of relationships among all beings in our universe, I begin to create a new myth. Couched in storylike language, although it may reflect all we know of dry scientific fact, the new myth tells of the ecology of the universe.
Sara Ebenreck

I still kept in mind a certain wonderful sunset which I witnessed when steamboating was new to me. A broad expanse of the river was turned to blood; in the middle distance the red hue brightened into gold, through which a solitary log came floating, black and conspicuous; in one place a long, slanting mark lay sparkling upon the water; in another the surface was broken by boiling, tumbling rings, that were as many-tinted as an opal; . . . high above the forest wall a clean-stemmed dead tree waved a single leafy bough that glowed like a flame in the unobstructed splendor that was flowing from the sun. . . . over the whole scene, far and near, the dissolving lights drifted steadily, enriching it every passing moment with new marvels of coloring.

Mark Twain, *Life on the Mississippi*

They are, perhaps, on a desperate quest for light. Is it beyond rational theory to assume that? What little we do know about the sun and ourselves indicates that there is a relationship more complex, more symbiotic than any we in this generation of science have yet discovered or even imagined. We have strong evidence that tells us the sun has an effect on our emotions, our bones, our sexual behavior, our resistance to cancer, our feelings of well-being or sense of contentment with life, our personal vitality, and our ability to do our jobs well, to live our careers to their fullest potential. Is it not logical, then, to assume that the people who must live with the long nights of the north would feel so suffocated by the darkness that they would set off across the Arctic in search of their star?

Many ancient cultures must have believed in the physiological and psychological benefits of the sun. Hopi Indians gathered for their meditations to "let the light enter the tops of their heads," and Bolivian Indians still practice a meditational routine that takes them to the edge of a high cliff where they sit "taking light" as they put it. Except for the Western world's, there is hardly a system of medicine on the globe which does not recognize the benefits of light and the sun's full spectrum. The Chinese consider color an integral part of diagnosis, and Mahayana Buddhists describe each energy center of the body as possessing certain colors of the spectrum.

There are strong signals that this nation's scientific community has at least gotten a start on catching up to the rest of the world's fascination with the sun's relationships to mankind. Dr. Kendric C. Smith, professor of radiobiology at Stanford University Medical School, said recently: "Sunlight is probably the most important single element of our environment, yet it has been largely ignored by the scientific community.

If, as surely we must someday, we move closer and closer to solar living as fossil fuels become more and more limited and nuclear energy more and more risky, we will almost certainly find ourselves also

Positions of Sun at hours of 10 & 11 A.M. 12 M. 1 & 2 P.M. on Dec. 28th 1910, from Nome, Alaska.

moving closer to better physical health, mental health, and more harmonious interrelationships with the human and natural community. Although research about the effects of the sun on each of us has barely begun, what is known is provocative, often startling, and indicates that our star is more of a presence in every aspect of our personal beings than we may ever have suspected. We are discovering that the sun is not only

The sun rose this morning at about 9:30 o'clock, but never really left the horizon. Huge and red and solemn, it rolled like a wheel along the Barrier edge for about two and a half hours, when the sunrise and sunset met at noon. For another two and a half hours it rolled along the horizon, gradually sinking past it until nothing was left but a blood-red incandescence. . . .

The coming of the polar night is not the spectacular rush that some imagine it to be. The day is not abruptly walled off; the night does not drop suddenly. Rather the effect is a gradual accumulation, like that of an infinitely prolonged tide. . . . The onlooker is not conscious of haste. On the contrary, he is sensible of something of incalculable importance being accomplished with timeless patience.

Admiral Richard E. Byrd, *Alone*

139

END OF THE LONG
ALASKAN NIGHT

At twelve-thirty, the tops of the spruce just below the cache catch golden fire. The blue-black winter look of spruce turns warm green in an instant. I thought the trees were black because they were frozen. It was the absence of sunlight.

We hurry outside, and the whole lights up.

The great bright glow moves surely, steadily, across the limitless snowfields. Reaches the woodpile behind the cabin and turns the stacked split logs a warm tawny yellow. Touches the back corner of the cabin roof and turns a long blue icicle into a glittering diamond dagger. The paths around the cabin fill up with brightness. The cabin is bathed in light.

Sun pours over our camp as, above the mountain peaks toward Jim Pup, the fiery globe itself appears in a blaze of hot gold. Sunlight now embraces all the frozen land. Sun warmth strikes our faces in a sudden gentle blow. We lift them to the light,

[*Continued on page 141*]

dazzled, blinded. Shout greetings to the sun as if welcoming an old friend who has been missing too long. Raise our arms and reach toward the source of warmth, toward the light which wraps around us, and everywhere, blue snow now gold-warm and splashed with a skyful of dancing brilliants.

A strange and joyful madness erupts in our camp. And we dance a lunatic, gleeful jig round and round the cabin with our boots making the music as they clump over frozen snow with drumming, brassy, crunchy sounds. We come around the cabin corner and meet our shadows, seen for the first time in seventy-seven days. Elongated and distorted by the low-in-the-sky midday sun, our shadows start a contest in mugging and jumping and dancing, and I wonder why elsewhere shadows have come to belong only to children!

Billie Wright, *Four Seasons North*

one of the most important cosmic forces, but a shaper of our physiology and psychology as well. We have, however, learned more about outer space than our own inner space. When every house, factory, and office relies directly on the sun for its light and energy, then, surely, we shall also verify what the ancients may have known tens of centuries ago: we are creatures who, like the birds and the flowers, resonate to the solar presence, and find our peace of mind, our very life-force, inextricably linked with the star whose rays sparked the first life on earth.

Eight years ago my family and I built what has become known as a "solar house." It is a place oriented to the sun—a structure that obtains the majority of its heat and light from the solar arc across these New England skies. I knew less about the sun when we built the place

In Minnesota, the whole of the state's demand for energy [in fact more than twice as much, according to a recent survey] could be met by annually harvesting the cattails that grow wild in most of that boggy state and using them to produce methane. . . . In the Arizona desert, the intense sunlight could be readily collected as heat or in the form of electricity from photovoltaic cells. In the Dakota Badlands where there is little vegetation and only moderate sunshine, there are strong winds, which could be used by windmills to generate electricity. In sum, there is no solar panacea . . . A national solar-energy-gathering system needs to be as varied as the country itself, reflecting its uneven climate, terrain, and vegetation.

Can the country eventually produce enough solar energy to meet the total national demand? In theory, the answer is yes, since the total amount of solar energy that falls on the land is hundreds of times as much as we need. But a practical answer is more difficult, for the demand requires that sufficient solar energy be produced in the appropriate forms. . . . The present demand for heat could be met by solar collectors covering about forty-four-hundredths of one per cent of the American land area, and the demand for electricity could be met by using twelve-hundredths of one per cent more for photovoltaic cells . . . In view of the fact that streets and roads take up about one per cent of the land area, this is not an insuperable problem. By prudently using available organic wastes, developing new sources such as seaweeds, and—most important—judiciously reorganizing agriculture so that it yields both food and solar fuels efficiently, the crucial need for liquid and gaseous fuels could also be met. The alcohol or other liquid fuels made from biomass available from agriculture could certainly provide for the expected needs of aviation, with enough left over to fuel heavy land-based vehicles that are not readily operated by electricity. The twenty quads of methane and hydrogen that would be produced could, if they were divided about equally between industry and the residential-and-commercial sector, take care of the demand for heat and electricity not readily met by local solar collectors, photovoltaic cells, and windmills

In sum, the route to solar future is open.

Barry Commoner, "The Solar Transition"

than I have since learned. Nevertheless, I knew after our first year in our new home that something important had changed in the lives of each family member.

The place has brought us a portion of serenity which none of us had previously known. That serenity is, at its core, an element of faith which springs from an increased awareness of the rhythms of the universe, of the turning of the earth on its axis, of the sun's recurrent rising and setting, the moon's waxing and waning, the tide's perigees and apogees, the wind's cruising of the compass, the circling of the seasons around us, and the frequent and irrefutable reminders that the cosmos is within us and without, with each of its intricate mechanisms in place, with each of its cycles—ours included—part of an order that is at once awesome and tranquil.

I am now more convinced than ever that more and more Americans will make the same sort of discovery as they move into their own solar living spaces that are certain to be part of the national future. I have learned, since we made our move, that more and more men of science are supporting the heartfelt reactions of myself and my family with objective observations of many aspects of the human condition when it is blessed by the sun. There is, as every investigator of the sun–personal relationship has made clear, much, much more to be learned. What we have already discovered, however, can only be classified as good. I have come to share Ott's observation: to keep well and happy, we will almost certainly put ourselves on sunlight diets some bright day in the future.

LXXXV

A LIGHT exists in spring
Not present on the year
At any period.
When March is scarcely here

A color stands abroad
On solitary hills
That science cannot overtake,
But human nature feels.

It waits upon the lawn;
It shows the furthest tree
Upon the furthest slope we
know;
It almost speaks to me.

Then, as horizons step,
Or noons report away,
Without the formula of sound,
It passes, and we stay.

Emily Dickinson

Live thou thy life beneath the making sun
Till Beauty, Truth, and Love in thee are one.
ROBERT BRIDGES

5   SUN LIVING

According to the United States Department of Energy, in 1970 there were a mere 24 homes in the nation specifically designed to utilize solar energy. By 1978, the official tally had reached 40,000 and was climbing faster than Washington bureaucrats could count. Projections of the current growth rate suggest that 1 of every 10 Americans will be living with some form of solar heating by 1985. Most of those 40,000 Americans building some form of solar assistance into their homes and/or offices were spurred to take action by the soaring price of conventional heating fuels.

It is the current uncertainty about supplies and the concurrent steep hike in prices that has made so many otherwise "practical" Americans turn toward the sun. They know little of the sun dances of the Cree, do not comprehend the magnificent mechanisms of the earth's orbit, and are unaware of the dimensions of the available solar energy that floods this planet and the space of our solar system.

Perhaps the greatest good accomplished by the foresighted and courageous individuals who have built or installed some form of solar energy system is the encouragement they give to others. There is a surprisingly insistent group of writers and speakers in the land who argue against solar energy for one reason or another (most frequently, they

William A. Shurcliff, Ph.D., a solar researcher and inventor with more than a dozen patents to his

credit. Shurcliff's work adds scientific and statistical weight to the contention that more and more Americans will soon discover ordinary and extraordinary solar benefits. Shurcliff, an honorary research associate in the physics department at Harvard University in Cambridge, Massachusetts, personally compiled an informal directory of the organizations and people involved in the solar heating of buildings. In 1977, however, he presented his third, and last, edition of the directory. For, in his own words, "There is too much going on under the sun to be put between the covers, of one, inclusive directory. It would have to be too big, too expensive. The only way we'll be able to keep track from now on is to talk about what's happening in specific regions, or with specific sorts of solar applications; or, say, commercial firms or private homes, or high-rise buildings."

represent a vested interest, like a public utility or a nuclear power investment group). Solar opponents insist that the sun is a pleasant presence in the sky, but can hardly be counted on for such staples as home heat, light, hot water, and fuel for the family's transportation. Few of us like to be considered fools, no matter how well meaning, so we tend to think of solar living as something done by members of the counterculture, those on the fringe of society who consider it an obligation to flaunt convention.

Most of the 40,000 people who are now on some sort of solar track are not nonconformists. They are, like one family I know, middle-class working people. My friends, the Davis family, installed a sunspace on the south wall of their modest home in a typical housing development. He manages a shoe store; she is a secretary; and they have two young children, a boy and a girl. Through a census taker's eyes, they are a classically typical American family. Their income, their education, their home, and their life-style have their counterparts in every community of every state in the nation.

How, then, did the Davis family come to make its rather unorthodox decision to tear out the south wall of a home they had only recently bought to install a relatively untried system which—on paper—would help heat their home in the winter, provide space for growing plants and vegetables, and (as it turned out) become a cheerful spot for their children to play? Theirs is the only home of some 45 in the development with any form of solar assistance; that fact alone made the decision a serious one, not to be made on a whim, or the desire to somehow be "different."

Among the reasons that helped the Davis family make its mind up were these: they lived in a community whhch is also the hometown of Cornerstones, the instruction headquarters for owner-builders who want an energy-efficient home that will allow its owners to live in harmony with natural presences. Because Cornerstones alumni account for many of the solar-oriented homes being built in the nation, and because most of

148

the early Cornerstones students came from nearby communities, there is a larger-than-typical number of solar homes within a day's drive of the Davis house. It was after they inspected those places that Holman and Roberta Davis were convinced that the adventure could have a happy ending. They were also motivated by the soaring costs of conventional ways of staying warm through Maine's invigorating winters, and by marginal tax incentives.

Their exploration of the unknown did lead them to a better land than they had left. The project, complete with expected delays and on-the-job improvisations—is a success. It is attractive looking, fits the house like a glove, was completed almost on time and quite within budget, and does the job it was intended to do: uses the sun as an energy source which warms and lights the house. Many neighbors have visited to inspect and admire the addition, and the contractor who took on the difficulties and the unexpected challenges of a first-time construction project now has at least three orders for duplicates of the Davis installation.

The Davis story, multiplied by millions, is the story of the advance of sun living in America. It has been a classic, grass-roots phenomenon. It has been done by trial and error, and by sharing the results of each trial, generously, freely.

In ten years of trying, the government has yet to build its proposed Solar Energy Research Institute—a facility budgeted in 1978 at $128 million. I am, I'm quite certain, in a small minority when I suggest that it's a good thing the government hasn't been able to coordinate its entry into the Solar Age. But, in my view, the evidence is on my side of the scales. As I leaf through the number of books that have been published cataloging and explaining hundreds of different sorts of solar-oriented homes in every state in the nation, I am convinced of the people's invincibility. It is governments that are so vulnerable, so hesitant, so careful, so behind the times. Which is, in my opinion, as it should be.

[*Continued on page 152*]

Of all the elements of the built environment, the home is distinguished as being a microcosm of the macrocosm it is set in. . . . By changing the house, we inevitably change society. What cannot be changed from the top may perhaps be changed from the bottom.

This implies a shift in domestic architecture from the Le Corbusien paradigm of the modern house as a "machine for living"—a direct extension of the fossil-fuel powered technological society—to the concept of architecture as an extension of the natural environment fueled by the same forces that drive the rest of the biosphere.

Sean Wellesley-Miller, "Toward a Symbiotic Architecture"

As so many people who live in solar houses remark, living with the sun registers on the psyche. "It is not just the financial savings. We grow more in awe of the tenuous hold our lives have on this small planet, more convinced that the sun renews us, in an almost religious way," says Junius Eddy. "It has made us profoundly grateful that the sun is up there, the center of our universe, warming us up and keeping us alive. That atavistic sense of the elements that early man knew and felt has become a part of our lives."
Design for a Limited Planet

Junius and Louise Eddy added a solar-heated wing to their house in Little Compton, Rhode Island.

A solar home is a workable way to stay warm as far north as Maine.

Thomason solar home, one of the first.

Heating a house to 68 degrees F. with electricity from a nuclear reactor in which fission at trillions of degrees heats the fuel to thousands of degrees is a gratuitous absurdity; . . . A single nuclear reactor, meticulously engineered, carefully tested, and thoughtfully sited a safe 150 million kilometers away–in fact, the sun itself–is quite enough.

Barbara Ward, Foreword to Amory B. Lovins, *Soft Energy Paths*

151

If we were to opt for the best renewable energy technologies, buildings could be engineered to take full advantage of their environments. More and more of the energy needed for heating and cooling would be derived directly from the sun. Using low-cost photovoltaics that convert sunlight directly into electricity, many buildings could eventually become energy self-sufficient. New jobs and professions would develop around the effort to exploit sunlight, and courts would be forced to consider the "right" of buildings owners not to have their sunshine blocked by neighboring structures.
Denis Hayes, *Rays of Hope*

"The business of government is government," and that implies the maintenance of the laws of the land, the protection of human rights and freedoms, and the management of the national defense and concomitant budgets. It does not imply a superparental presence that tells us what we must do about everything, and when, and how.

Certainly the builders of sunspaces, of sun machinery or sun devices, sun plumbing, sun hardware, sun houses, and sun schools were not guided to their goals by their government. Quite the opposite. Tens of thousands of others, who might have moved closer to the sun, have been discouraged by ordinances like those which prohibit clotheslines, or insist that the flush toilet and the septic tank are the only proper systems for transporting and treating human wastes and dishwater.

Those who have not been discouraged, those who have cleared every nit-picking legal hurdle, those who have the self-confidence to believe in their own solar visions and who will not rest until those visions are realized, those 40,000 Americans—0.2 percent of the population—are the men and women who will lead the other 99.8 percent of the citizens into the Solar Age. Sometime after the next century has gotten a good grip on our calendars, someday when the nation has become a solar culture, a proper solar memorial should be created someplace honoring the individuals of the sixties and seventies—the solar pioneers who listened only to their common sense, their instincts, their wonderfully human desires to find a better way, no matter how many mistakes they had to endure in the process.

They tried everything, and they made every kind of mistake, and scored several kinds of successes. Empty milk containers, full beer bottles, cinders from the coal-burning furnaces of the cities, mud, adobe, tar paper, and empty oil drums were on the list of the building materials the pioneers worked with and built with. They created mystically elaborate designs and startlingly simple ones. They utilized new devices to collect, store, and distribute solar energy, thus creating what are known

as "active" solar systems. And they architected their dwellings so that the sun poured into their spaces as it moved through the sky, filling the homes with heat and light, and a sense of vast accomplishment. Thus were "passive" solar systems born, and became a way of utilizing the sun without dependence on gadgets that depended on materials, that depended on availability, that depended on sophisticated techniques.

The solar pioneers took 200-year-old Yankee homes, built by their pioneer ancestors, and with rooftop panels, or double-glazed, south-facing windows, they converted them to homes that used the short winter sun of the Northeast as a heat and energy source of significant magnitude. Or they built new places, from the ground up. Places soared into the sky, or huddled against hillsides; they were fashioned from wood that had been salvaged from barns destroyed to make room for new highways, or they were literally dug from the faces of sandstone cliffs. They had water tanks on the roof, transparent plastic in the skylights, cobblestones in the basement (to hold the sun's heat) and walls within walls within walls. They were half greenhouse, half living space; or half fish pond, half bedroom. There was no design that was not tried. Trailers were converted, tepees were insulated, domes were set atop domes and yurts atop glass yurts.

Only the yearning for the sun was uniform, only the determination of the builder and/or designer to demonstrate a kind of independence, a belief in the beneficence of the solar presence, a conviction that the high-energy, inefficient disposal of fossil fuels was somehow a false logic that would inevitably be repudiated. Only these qualities and a vigorous self-confidence and a wondrous ingenuity were shared by the solar pioneers. And working in their separate but shared ways, they gave the Davis family, and hundreds of thousands of their fellow Americans, some notion of the roads that should be followed, the solar paths that should be taken, and the solar thickets that are best avoided. They built high-cost palaces, middle-range homes, and low-budget and no-budget homes that

Charles Greeley Abbott, early solar experimenter at the Smithsonian Institution in Washington, D.C., stands with his model solar cooker.

Why do solar buildings have some magical appeal? We are not particularly interested in a building when it is heated and powered by natural gas, bunker oil #2, or hydroelectricity. These energy forms seem to matter not a twit in conception of a building so long as it works. But, a building based on the sun is blessed. It implies a special conception—the promise of an almost timeless relationship to the cosmos and the potential of an infinite continuum of vital energy . . . Solar architecture is building conceived in the womb of natural processes. It is design oriented to nature, both earthly and heavenly.

Jeffrey Cook, "Concepts of Solar Architecture, Archeoastronomy and Autonomy"

Solar panel system supplies more than 60 percent of the heating and cooling for this Atlanta, Georgia, school.

154

Farallones Institute, Occidental, California.

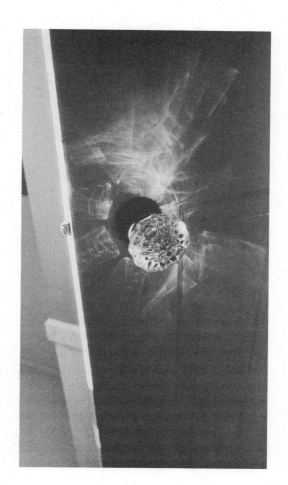

still stand as a monument to diversity and the belief that anything can be done if it's wanted to be done seriously enough.

The process was not with its pain, its failures, and its traumas. But it was always more than a fad: it was, and is, a bona fide movement, a social transition of impressive magnitude and rare purity. It is happening with solar collectors, solar reflectors, solar systems, solar storage, solar stoves, solar furnaces, solar power, solar engineering, solar design, solar siting, and hundreds of solar stories from each of the solar pioneers about the unexpected and often unexplained benefits of solar living.

In the United States now, none of us is far from a domestic solar installation. We can go visit, see, touch, ask about, and copy for ourselves—or improve on—a solar installation built by the people and for the people. This could not have happened a few years ago.

But if you could locate the original 24 adventurers and put them in a room with the subsequent 40,000 solar explorers who have

[*Continued on page 158*]

An artist's conception of a 10-megawatt solar thermal plant. Movable mirrors reflect solar heat, focusing it on a boiler at the top of the solar tower where the high-pressure steam is pumped down to drive electrical generators. Government contractors are testing component parts, aiming for a working thermal plant by the early 1980s.

Das Brennfpiel in der Mittagsfonne.

Arthur Katz is one American who would raise a doubtful eyebrow at circumstantial evidence of a burgeoning solar society. Katz spent more than $10,000 in a year-long legal battle with the Mamaroneck, New York, zoning board. At stake was his right to install a $2,700 set of solar panels on the roof of his home. The panels, whose design was innovative enough to win Katz a $400 grant from the United States Department of Housing and Urban Development, are now in place, but many of his neighbors remain skeptical.

The sun . . . A cat dozing in the sunny doorway of a barn knows all about it. Why can't man learn?
E. B. White

since followed the same course, you would find that the entire group had similar notes to compare—especially when they began sharing their thoughts about some of the unexpected and unplanned benefits of sun living. A kind of cosmic shift occurs in the consciousness of Americans when they are transported from the world of fossil fuels in which they were born and raised. As they enter the fringes of the solar environment, they discover much more than the comfort of radiant heat, or the view from south-facing windows. What they had lost, what so many of us have lost during the petroleum age, is a feeling of being in touch with the universe, of comprehending the planets, of sensing each day that there is a magnificent and mysterious order to every moment, that a procession of which we are a part is making its way with sureness and dignity along a wide road that leads—we are sure—toward an age of increasing human fulfillment and a time of peace.

Sun living does this. There is too much evidence to argue otherwise. Talk with the 40,000. Every one of them will have a story about a serenity they have come to know, a security they had never expected. "I never feel alone anymore. . . . I have learned to be at peace with myself and my family. . . . I have never been afraid in this house. . . .You would be amazed at how my plants grow. . . . Our pets have never been healthier. . . . The children love it. . . . We have never been a happier family. . . ."

There are 40,000 statements like these, endorsements of a way of living that began as an effort to save money, to avoid the insecurity of uncertain fuel supplies, and evolved to new and revelatory dimensions that are still not fully explained or experienced. We are all beginning our solar explorations. Just as Columbus and those who followed him had no possibility of grasping the scope of the nation that stretched westward from where they stood on the easternmost shore, so are we standing on the outer reaches of an age whose promises and potential are, quite simply, not yet able to be predicted or defined.

This Campbell Soup facility in Sacramento heats
its water with solar panels.

Two hundred years from now, when the petroleum age will seem as remote as the Stone Age, our heirs will look back at this time with as much curiosity as we probe the first civilized meanderings of the Neanderthals. We wonder at times how those primitives were able to avoid perceiving the many benefits of fire. Why, we wonder, did they fail to grasp the energy inherent in boiling water? How could they avoid seeing the effect fire had on metal, on clay, or on the foods that were taken from the trees of the forest or the animals of the plains? The primitives collected wood, struck flint to iron or waited for lightning, and painstakingly, carefully, with infinite patience built a fire, and then utilized it only as a source of warmth, a way of increasing the comforts, or easing the discomforts, of their cave dwellings. How could they have failed to see that fire could do so much more than that? We ask those questions now just as innocently as citizens of the twenty-second century will inquire about our failures to recognize the solar potential that is there before us.

As an optimistic futurist, R. Buckminster Fuller compares humanity to a tiny bird pecking its way out of the shell of its gestation, needing to learn how to operate on an entirely new basis.

"It is possible for humanity to prosper while employing only the natural energy income of wind, tide, sun, gravity as water power and electromagnetics of temperature differential," Fuller writes in The Futurists. *But ". . . We are going to have to find ways of organizing ourselves cooperatively, sanely, scientifically, harmonically, and in regenerative spontaneity with the rest of humanity around the earth." In short, "Henceforth we can survive only by learning to operate in our universe in a very different way."*

159

Making electrical power from the sun's light for transistor amplifiers: a Bell Telephone's solar battery, linked also to a storage battery.

Solar power operates the cooling system in this Holiday Inn, Virgin Islands.

So be it. We have struck flint to steel. We have, at least, made a beginning. There are more than 40,000 homeowners among us who have come to know more about intricate celestial relationships merely by observing the changing shadows on their floors. As dwellers of the temperate zone, each becomes a Copernicus or Galileo, discovering the essentials of the sun's relationships and its annual movements—complete with

the multidimensional variations. But no telescopes are needed, no Copernican dedication is demanded. Those who live with the sun are offered cosmic insights if they calculate the overhang of their southerly roofline, or mark the sun's furthest reach at the winter solstice. It is inherent in solar living that each becomes re-admitted to the natural, that we experience a sensory renaissance that extends our understanding of the place we live. We are not only given a novel (for us) method of warming our bedroom, we are granted the materials for a restoration of faith.

We come to know a different sort of light—an interior illumination that changes tones as clouds move across the sky, that alters intensities as the earth tips on its seasonal axis, that signals the imminence of a storm, that forecasts its violence, that gives the earliest signal of its passage. We can anticipate how the wind will turn through the compass, we can witness the magical diversity of snowflakes falling, and we can track the migrations of the monarch butterfly and the black-bellied plover as they respond to seasonal impulses we cannot see. Each equinox and solstice becomes an event as palpable as nightfall, as welcome as the dawn. The phases of the moon become as well known as our own, the stars are tracked through their journeys. Every presence above us and beyond our solar windows is re-admitted to our existence. We find new friends in a world that has always been with us, but which we had never seen.

We are warmed, we are informed of the seasons, we are welcomed to our natural world, we are moved to cosmic comprehensions, we are blessed with a new security and contentment, we are made aware of our ability to be self-sufficient, we are no longer alone, our children, our plants, our pets, our lives are enhanced, we are relieved of the anxieties of resource scarcity, and, in what is surely an incredible American fillip, we save money while our oil-burning neighbors must spend. The sun gives us each of the gifts, and gives them with ultimate generosity, expecting nothing in return. We are so taken with what we have discovered on this

Sunlight leaves an earth unravished, husbanded, renewed. It leaves a people unmutated, convivial, even illuminated. Above all, it respects the limits that are always with us on a little planet: the delicate fragility of life, the imperfection of human societies, and the frailty of the human design. We can still choose to live lightly, to live with light, and so choose life itself—by capturing the hope left waiting at the bottom of Pandora's box.
Amory B. Lovins, *Soft Energy Paths*

Light is the lion that comes down to drink.
Wallace Stevens, "The Glass of Water"

There are many projects which demonstrate the solar reality to industrial and urban doubters. By optically squeezing sunlight, inventor Lynn Northrup gets higher temperatures from his commercially oriented, rooftop solar collector than most solar-heating systems now on the market. Northrup, who is head of a Hutchins, Texas, company that makes heating and cooling equipment, has combined lenses of extruded plastic—a lightweight, easily produced magnifier—with water-filled tubes two inches in diameter. Built to follow the solar arc, the collectors supply a New Jersey Burger King with all the hot water it needs at temperatures of 220°F. Northrup is confident the system can be further developed and will produce water at 500°F. for a variety of industrial applications.

The Ishibashi Solar House, designed and built by the director of Yazaki Corporation, was the company's second solar home.

first step of our solar exploration that we imagine there is little or nothing left to be discovered. Like the Neanderthal, thrilled beyond explaining when he discovered fire's heat, we are so excited by re-discovery of what the Roman architect Faventinus had learned in the fourth century that we are quite content to stay where we are, to camp here, at the edge of the easternmost solar shore, even though an entire solar continent waits to be explored.

Some solar scouts have been dispatched; more are on their way, pressing forward into unmarked territory. As ignorant as Columbus was of the Mississippi's might, the solar inventors and tinkerers can make

more of a difference in shaping the world's future than the explorers who found the Pacific made to the future of the United States. With more solar discoveries—and they are surely waiting to be made—the value of the remaining oil reserves becomes more rational, able to be discussed, subject to reason. With enough solar discovery, what little oil remains becomes quite irrelevant.

This is the importance of the solar scouts—the men and women who press beyond the easy landfalls that have been made, the individuals who are already aware that we haven't done much more in the twentieth century than Faventinus did in the fourth. Not only are they making it simpler and more tempting for the next 40,000 solar-space builders to follow them, they are discovering the technologies and logging the insights which can convert that number to 400,000, and later, 4,000,000.

The results are as varied, as imaginative, and sometimes as impractical, as the several hundred designs and schemes developed during the first years of the twentieth century, when every inventor worth his title experimented with ways to utilize the internal-combustion engine for personal transportation. If you could visit the United States Patent Office, or the libraries of those who collect automobile memorabilia, you would find it difficult to resist laughing at some of the drawings—sketches for machines that could move humans from one place to another.

That was 50 or 60 years ago, not even an actuarial lifetime, yet now the 2,000 or so American companies that were organized to manufacture and market one sort of car or another have become 4. No one doubts the efficacy of the automobile. It has become as much a part of living in this nation as the home and family. Ironically, it is the grip the automobile has on the culture that is helping to hinder solar progress; many people are concerned that the transition from petroleum will mean the loss of "freedom" that the car has become. Yet the machine did not exist 75 years ago.

There lives the dearest freshness
deep down things;
And though the last lights off
the black West went
Oh, morning, at the brown brink
eastward, springs.
Gerard Manley Hopkins

The adventure of the sun is the great natural drama by which we live, and not to have joy in it and awe of it, not to share in it, is to close a dull door on nature's sustaining and poetic spirit.
Henry Beston, "Midwinter," *The Outermost House*

Printing press operated with solar energy at a
Paris demonstration in 1882.

164

Mouchot's Solar Engine.

Michigan project workers designed, constructed, and operated two generations of commercial-size solar greenhouses through three Michigan winters. The final modification of the effort is an A-frame solar garden some 6,000 square feet large, built primarily of wood and utilizing transparent fiberglass as its sun source. Well insulated, the garden was planted with lettuce, broccoli, radishes, and other hardy crops in October. The crops matured and were harvested, as were second crops of lettuce and spinach, picked in April. In spite of several days when outdoor temperatures registered below-zero levels, no additional heat was needed or used. Only the sun, working in harmony with the structure, was necessary to allow farming through the winter months in the frostbelt.

Sun-treader, life and light be thine for ever.
Robert Browning, "Pauline"

That is a thought to keep in mind when you examine the number of solar-oriented systems that have been created in the past few years. Yes, they are a mixed collection, yes, many of them have the look of gadgetry, and, yes, some of them even appear comic—the work of the stereotypical "crazy" inventor. Some of the 2,000 different designs for automobiles looked precisely the same several decades ago, yet we are

FIG. 1

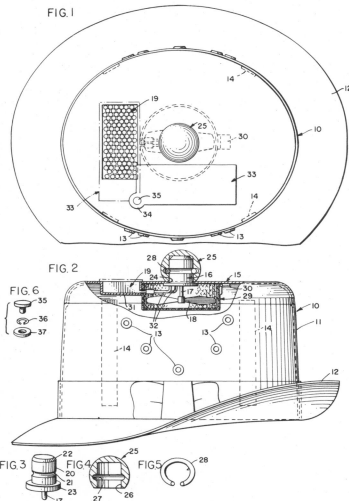

FIG. 2

FIG. 6

FIG. 3   FIG. 4   FIG. 5

An ingenious help for the hot-headed man. A solar cell on a plate runs a motor which powers a fan in the hat. Cooling the top of the head will have a cooling effect on the entire person, the inventor says.

Early inventors came up with some elaborate if not ingenious schemes for preserving foods.

now a nation on wheels. Several decades from now, we shall be a solar-heated nation, and several decades later, a solar-powered one.

Henry Ford and Walter Chrysler did not begin their work at assembly lines in Detroit; they began in their barns, basements, and backyards, just as Studebaker and Dodge also began. And it is in the backyards, basements, and garages (now the barn's equivalent) that today's solar inventions and insights are being assembled. Who can doubt, given the past, that the future will not evolve in much the same ways? Looking ahead a few years—say 20 or 30—I can see homes that

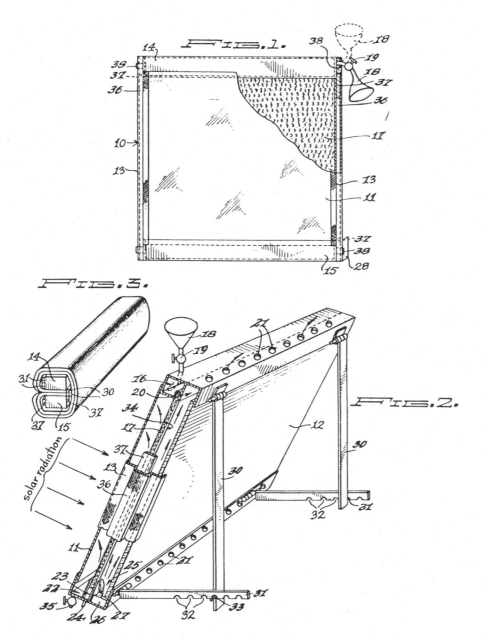

Downed planes and torpedoed ships at sea during World War II led to the invention of compact devices for distilling fresh water from salty seawater—the edge of life for a crew in a life raft.

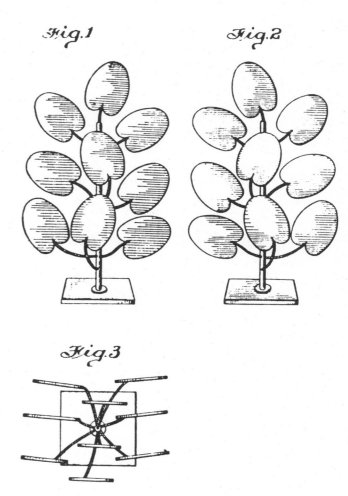

Not all solar applications were aimed at the practical. Here, a hopeful inventor sought to capture the leisure market with this reflecting unit for sunbathing.

167

The best approach (to a solar society) may be to "let a thousand flowers bloom" and then to cultivate the most promising varieties. We have a society where hundreds of thousands of citizens are skilled in science and technology. It may take a solar technological revolution to remind us once again that the true strength of such a society is revealed when its members are given the opportunity to show what they can do.

Robert Williams, *Toward a Solar Civilization*

appear much the same as those being built today. They'll have a bit more glass on the south side, just as apartment buildings, office buildings, schools, and other public institutional buildings will also. Solar heating and cooling systems will be built into that glass and regulated by a central thermostat.

In summer, the homes and buildings will be solar-conditioned; in winter, they can be solar-heated to precise degrees of comfort. There will be no bulky apparatus, no cellars filled with rocks, no plastic containers or recycled oil drums painted black, no cumbersome roof panels and light concentrators, trickle plates, copper piping, ducting and fans. There will be windows, almost as we now know them, structured to absorb and store and release the sun's light and heat in harmony with the needs of the homeowner.

What the home inventors and the tinkerers are doing now, in these first decades of solar technology, is capturing the sun's energies with mass, whether it be the dark floor, the containers of water, the "Trombe" rock walls, the piped water, the crushed cinder block, or merely the interior mass of the structure itself. It is this principle which is utilized in each of the current solar-heating systems.

Capturing the sun's warmth, its energy, is the key to solar living. Methods of doing that now correspond to the methods used when gasoline engines were mounted on carriages and buggies—the vehicle used came from a previous era. When solar living becomes a reality for 200 million Americans (and for the rest of the world), it will happen because an integrated set of solar systems will be created. Structural design and structural heating, energizing, and lighting will be keyed from the start to a solar source. There is nothing beyond the realm of known technology that could prevent solar convertors from being built into windows; what we must do is quite thoroughly revise our notions about what a window must do.

It would be·interesting to see the results of a contest among the

This French solar furnace, located in the Pyrenees, uses mirrors (foreground) to reflect sunlight onto the curved surface of the office building in the background. That, in turn, focuses the sunlight on an opening in the tower at center. There, temperatures of 7,000°F. can be produced—enough heat to melt any known material.

world's scientific community. Suppose individual prizes of $10 million each were offered to the individual, not the corporation, who could design a solar window, a self-contained, self-regulating unit which could utilize the sun's full spectrum. With proper materials—and they are surely within the realm of reality—and proper design, the solar windows could collect the sun's energy during the day, meter it to insure an even flow, diffuse and store surplus heat, and, in a later generation, use the surplus balance as an energy source able to be converted to electricity to power household appliances.

Corporations and governments have spent 10 times $10 million, and produced next to nothing. Given the inventive drive, the creativ-

[Continued on page 173]

SOLAR INSIGHT
FROM SOCRATES

His dictum about houses was a lesson in the art of building houses as they ought to be. He approached the problem thus. . . . "Is it pleasant," he asked, "to have it cool in summer and warm in winter?" And when they agreed with this . . . "Now in houses with a southern aspect the sun's rays penetrate into the porticos in winter, but in summer the path of the sun is right over our heads and above the roof so that there is shade. If then, this is the best arrangement, we should build the south side loftier to get the winter sun and the north side lower to keep out the cold winds."

Xenophon, *Memorabilia*

Southwestern native American pueblos were designed to integrate the lifes of the dwellers with the cycles of nature. Pueblo Bonito in New Mexico was constructed as a cup-shaped solar collector. Its design admitted the winter sun for heating and repelled the summer sun to keep the structure cool. Specific illumination points marked seasonal changes.

Design for interior of the Bank of England, 1798, by John Soane.

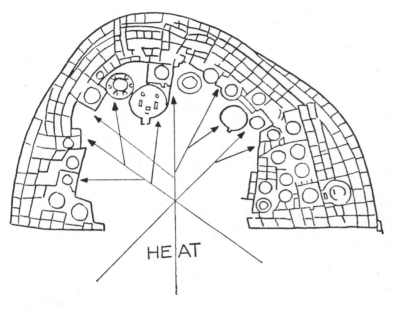

HEAT

Pueblo Bonito.

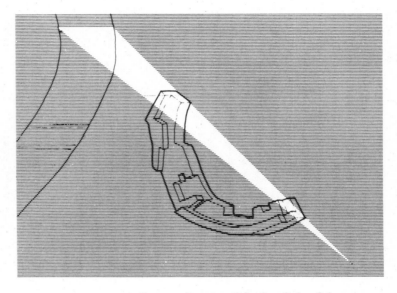

Above, a diagram of the first light of the summer solstice on the edges of Pueblo Bonito.

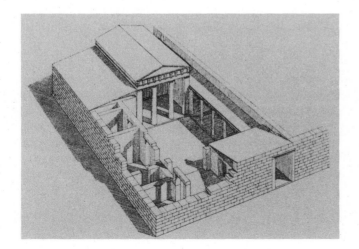

Reconstruction of a classical Greek home from excavation of the city of Priene by Theodore Wiegand. The rooms behind the portico faced south onto the courtyard.

171

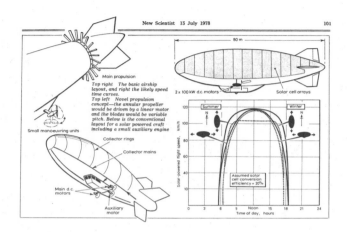

New Scientist 13 July 1978   101

Main propulsion

Top right The basic airship layout, and right the likely speed time curves.
Top left Novel propulsion concept—the annular propeller would be driven by a linear motor and the blades would be variable pitch. Below is the conventional layout for a solar powered craft including a small auxiliary engine

Small manoeuvring units

80 m

2 x 100 kW d.c. motors   Solar cell arrays

Collector rings
Collector mains

Main d.c. motors
Auxiliary motor

Assumed solar cell conversion efficiency = 20%

Summer   Winter

Solar-powered flight speed, km/h

Time of day, hours

Scientists are studying the potential for solar airships to carry large loads in areas of the world that lack established roads or railways. Since many of those areas have high levels of solar radiation, a solar airship may soon be more than a flight of fancy.

Twenty-three centuries ago, Archimedes knew the principle of concentrating heat energy by focusing the sun's rays on a point of concentration. According to tradition, he set the wooden ships of the attacking Romans ablaze in Syracuse harbor by using mirrors to direct the sun's rays at their hulls.

ity of today's solar tinkerers, that money could surely be better spent sparking and financing American ingenuity. The problem with solar living is not the sun, it is our culture, its economic system and its politics. Those who have grown with the Industrial Age have seen centralization become the key to making a profit. That's why there are just 2 viable automobile companies left of the 2,000 that existed 60 years ago. That's why nuclear power proponents are so unwilling to abandon their capital-intensive, highly centralized energy plans. That's why no major companies have yet participated seriously in any aspect of our solar future. They cannot foresee how they can maintain control of the marketplace, how they can make money as massively as they are currently making it.

Home solar energy systems do not, by their very nature, lend themselves to centralization. Like sunlight itself, they are diffuse, independent of each other, tailored to individual needs. The sun cannot be metered. One family's solar needs will be supplied as generously as another's, one housing project home will get just as much solar exposure as another, and the energy does not come through wires or pipelines, but from the sky above.

Existing utilities operate on quite different principles. They supply energy in terms of how much the consumer can pay; energy consumption relates to earning power rather than home design or location. The fossil-fuel energy system is a product of growth economics and the politics of capital growth which the system has fostered. Solar systems, on the other hand, will create equilibrium economics, a more serene financial environment in which the individual counts for as much as the corporation. With the Solar Age will come a time of economic balance rather than incremental annual growth.

It is because a great many people are charged with anxiety about the transition to such a system that the advance of solar living has been such a halting, diverse process implemented by inventors and tinkerers, not corporations or the government. Very few company presi-

The sun: grows $100 billion worth of wheat, corn, etc., each year; grows five trillion trees and shrubs, 17 trillion flowers; provides daytime illumination for 200 million people; melts 10 trillion tons of snow and ice each spring; vaporizes six trillion tons of water that result in rainfall supplying three million acres of farmland, 1,000 rivers; warms five trillion tons of coastal water in which 10 million people swim; produces winds that drive 100,000 windmills, propel 700,000 sailboats, lift 200,000 kites, dry one billion freshly laundered sheets, and purge smelly air from one million city streets.

William A. Shurcliff, *New Inventions in Low-Cost Solar Heating*

If we switch from looking at building-by-building applications of solar energy to seeing community land areas as a whole, we'll find that cities have much greater solar potential, says environmental planner Ron Cornman, author of a study, "The Potential of Solar Energy in the City of Baltimore by the Year 2000." Seeing the city as one vast expanse of collector surface, mostly made of roofs, leads us past the impasse presented by looking at separate buildings, some of which have much more roof-collector space than they need, some of which have less. The distribution of energy from the "surplus" buildings to the "deficit" buildings can be accomplished through the existing electrical grid for solar-photovoltaic electricity and through storage for solar-thermal heat.

[Continued on page 175]

dents are willing to abandon a profitable past for what they perceive to be an uncertain future. Quite the contrary. Some of them are actively resisting a solar future; others are attempting to support and test centralized solar systems like the billion-dollar experiment in the southwestern desert.

On the other hand, as inventors—with or without (almost certainly without) a $10 million prize—get closer each year to producing the self-regulating, self-storing solar window for home, office, school, and factory, centralized corporations now in the business of making money from energy consumers may see the writing on the glass and begin crash projects to keep energy centralized and metered. We shall have to wait and see, but we shall not have to wait long. As I write, the price of home heating oil in New England is just double what it was a year ago. These trends are certain to continue, at least until petroleum reaches its "true" market value. That would mean yet another doubling of home heating costs. In addition to the shock value of such increases, they also put a major crimp in the so-called payback schedules that have been so popular with solar opponents.

It is typically American to compute the value of the sun in payback numbers, but comparing the cost of solar energy hardware to the equivalent fossil-fuel mechanisms is a comparison without logic. We all, in the course of our immediate future, will have to use less fossil fuel because there will be less of it. And, in just one tick on the cosmic clock, it will be gone. Then computations of payback for solar installations will not only be illogical, but absurd; the sun will be our energy source. We shall all be living in a solar society. And while there may still be debates about which eras were "better" or "worse," the days of fossil fuels will be over.

Knowing this, all of us should be more grateful than many of us are for the work being done by the men and women working in homes and basements and garages. The solar tinkerers are fabricators of our

And if we move from our "right now" perspective of heating in winter and cooling in summer to a longer vision of seasonal solar rhythms, we can work on ways to store excess heat in the summer for release during cold months and vice versa for excess winter cold.

The trick in anything you do, thinks Cornman, is learning how nature does it. Like a forest with its unity of structure, its diversity of collectors and distributors and its enormous capacity for smoothing and storage, a city could learn to utilize the powers of solar energy. And in large-scale solar planning, Ron says, the first stage should be to see this from the sun's perspective, from up there rather than from our human perspective way down here on the ground.

George Rebh, *Editorial Research Reports*

Cities can be solarized as easily as the country.

175

What role do solar greenhouses have in bringing us a future that is richer than the present? . . . For many people, an attached solar greenhouse is an investment of sorts. It reduces energy and grocery bills, and if properly built, appreciates in value each year along with the rest of the house. For the hedonist, a greenhouse provides a luxurious space in which to bask even in the dead of winter. However, solar greenhouses represent more than economic rationality or sensual gratification. They are symbols of our vision for a smooth transition to a more civilized future; they are cultural archetypes for a sane and benign society. . . . Greenhouses are often among the most complex of biological environments, life support systems under the direct management of the people they serve. Their successful operation comes as the result of cooperation with the world we live in, not domination of it. As such, they can demand a depth of understanding akin to that required in dealing with society's most complex paradoxes. Therefore, although millions of solar

[Continued on page 177]

future. The unit they so proudly put atop their roof, or attach to their home's solar side, or build into their living room walls may be the one that holds the secrets of the solar window. We don't know which one, or how, or why. What we do know is that if the researchers and the experimenters and the constructors stay with their solar search, they will, certainly, find their solar answer. When they do, it will become our solar solution as well. Because it evolved from the free market of the individual, it will be designed for the individual and the spaces they live in.

greenhouses may supply only a fraction of our nation's total energy, they can serve as vehicles both for individuals and for the nation in the transition to a better future.
Bruce Anderson, executive editor, *Solar Age*

Meanwhile, just as there are signs that the solar tinkerers and inventors are bringing us closer to solar-home life than we may realize, other changes and developments are taking place that are moving other aspects of our lives into the sun. Of these developments, the rise of the solar greenhouse is perhaps the most distinct, as well as the most closely connected to the home.

"Solar greenhouse" may seem a redundant term to many of you. But in subsequent days of the Industrial Age, when petroleum and

In Canton, Mississippi, this rooftop solar system
provides 40 percent of the heating for its lumber
kiln.

coal were not only dirt cheap but readily available, greenhouses became
as dependent on supplemental heat sources as they were on the sun. They
were proper greenhouses only in the sense that plants grew under their
glass-paned roofs because of the wondrous interaction of the sun's rays
and the chloroplasts in every leaf. The heat generated was a solar surplus,
not to be counted on.

All that has changed, of course. As the cost of conventional
energy has soared, as heating fuels have become less taken for granted,

and as the urge to be less dependent and more self-sufficient has gained a kind of cultural popularity, so too have greenhouses. They are, if they are properly constructed and carefully insulated, able to operate, even in very cold latitudes, solely on the sun's daily infusion. Built as part of a home, they can supply solar heat, and a warm place to sit amid your own greenery. But they can also be used as indoor gardens, and it is this application which is most applauded by greenhouse owners. There is nothing quite so satisfying (I'm told) as collecting your own, homegrown lettuce, tomatoes and other such succulents while, just beyond the glass, a blizzard rages and many of your neighbors can't get to the supermarket to buy vegetables that have been picked days ago and shipped thousands of miles.

Just as every aspect of the Solar Age is more sensitive, complex, and subtle than it seems at first glance, the dynamics of greenhouses, atriums, sunspaces, conservatories, growholes—or whatever names they go by—are proving a surprise to many of the people who design and build them, and/or have added them to their living space. They are turning out to be among the most complex of biological environments; they become life-support systems for the people they serve in ways that are only beginning to be understood and documented.

They do more than provide space for vegetables and other plants to grow during the months of frost, and they do more than supply supplementary solar warmth to their adjoining dwellings. There are those who claim the ongoing process of photosynthesis invigorates and purifies the interior air they breathe, that it maintains a balance of relative humidity which enhances comfort and helps prevent scratchy throats and other symptoms of forced hot-air heat, or any overheated home.

Just how is not fully understood—yet. Someday soon it will be; someday very soon, solar greenhouses will no longer be considered as "exotic and expensive" as they are in the minds of many of the editors of the national home publications and the "Living" sections of many daily

Daniel and Sandra Newman built an octagonal, earth-covered, Navajo-like hogan warmed by an attached solar greenhouse near Cerrillos, New Mexico. "We must reinvent this kind of elemental structure for our own social and psychological survival. . . . It's renewing our sense of place in the world." *Design for a Limited Planet*

newspapers. They will become as much a part of every average home as kitchens are currently. Indeed, they should become an annex to every kitchen. When they do, an entire range of appliances and implements will be designed and marketed to help homeowners maintain and manage their new addition.

I once visited Scott and Helen Nearing at their home in Harborside, Maine. The work done and the books written by these two veteran articulators of organic farming and postindustrial living have made them national figures; the details of their contributions are so well known they need not be repeated here. I marveled at Scott's gardens, his massive and productive mulch and compost bins, the plumpness of his blueberries, and the carefully planned fertility of his farm ponds. But, as one who has lived through my share of Maine winters, I was most impressed with the solar greenhouse Scott showed me.

Built into the earth, with the steep excavation on the north wall, sloping banks on the east and west, and a glassed, angled front on the south, the basic, almost primitive structure cost very little to build. Scott, then in his late eighties, had piled some fairly large stones against the earthen north wall—these stored some of the solar warmth that spilled through the southerly windows.

"We grow vegetables in here all winter," Scott said, most matter-of-factly. "They never freeze."

I was properly impressed. The Nearings live some 150 miles north of me, and I know the temperatures can sometimes reach −20°F. I commented on the structure's remarkable properties.

"Nothing remarkable about it. We've had one of these for 60 years. Had one in Vermont, and it was colder there."

As I remember, Scott never once called the structure a solar greenhouse or a sunspace. It was merely a place where a vegetarian family could harvest their own vegetables through the cold and snowy months of a Maine winter. I have since learned that the Indians who

inhabited the north country before either myself or the Nearings also utilized pit greenhouses, as did some of the settlers who learned a great deal about survival from the Americans who were here first.

I'm certain the United States Department of Energy did not count the Nearings when they made their first survey of solar families a decade or so ago—the one that turned up 24 American solar homes. Yet Scott and Helen, the Iroquois, the Algonquin, and the Cherokee had been living in harmony with their own growing spaces for a good many winters before the Department of Energy was created. Their "solar greenhouses" were neither exotic nor expensive. On the contrary, they cost almost nothing, and the calories they converted often meant the difference between survival and starvation.

Besides greenhouses, small, diversified solar projects have been tested by individual farmers around the nation. In northwestern Minnesota, grain farmer Jerry Buckingham—who came to farming as an electrical engineer with a degree from Purdue University and five years' experience with Honeywell, Inc.—has designed and constructed solar grain dryers. His figures indicate his crops—wheat, barley, rye, and sunflowers—can be dried by the sun for one or two cents a bushel. He has no dairy herd or large herds of any kind, but suggests that there is enough solar heat left over after the grain has been dried to heat a milk shed or a farrowing barn.

Utilizing essentially the same principles for a different purpose, another solar tinkerer has put the sun and grain to work to distill alcohol. Just as it does with the fossil-fueled conventional energy sources, the sun starts the mash bubbling, alcohol condenses from the vapor, and, in this case, is collected and bottled quite legally under the supervision of the United States Bureau of Alcohol, Tobacco and Firearms which oversees the experimental work being done by farmer Duane Crombie of Webster, Minnesota. Farmer Crombie feeds the alcohol, properly denatured to make it unfit for human consumption, to his truck and tractors. Using the

The Don Chambers family of Dolan Springs, Arizona, used their solar greenhouse as a place to raise baby chicks fresh from the incubator. As the chicks grew in the sun's warmth, they shared some of the vegetables that grew along with them. When spring brought warmer weather, the flock was mature enough to be moved outdoors into a larger pen.

Inside the Nearings' greenhouse.

181

Kaplan Industries built this facility to convert manure into methane gas. The facility provides 88 percent of the energy for the processing plant, producing steam for heat, hot water and cleanup. The methane system replaces most of the 6,250 gallons of fuel oil previously required each week by the operation.

Gasohol, gas mixed with alcohol derived from plants, is viewed by some as a partial answer to our dwindling energy supplies.

solar still to extract a renewable fuel makes this particular energy circle a complete solar triumph; it is the sun that energized the corn's growth, and then powers the still that extracts the alcohol from the mash. Emulated on a national scale, every small farmer could use the same basic formula to manufacture the fuel needed by his farm machinery.

Taken another step further, the same solar plan could be used by many rural area homeowners looking for supplemental energy sources to power the family vehicle, lawn mowers, and electrical generators.

All that needs to be done, really, to create home and farm uses for solar power is to analyze what power appliances already exist. A company in Kansas City, Missouri, is already doing this and has produced, among its products, a sun-powered electric fence. The fence charger, manufactured by the Parker McCrory Manufacturing Company, charges up to 25 miles of fence. A solar panel charges eight volts into a six-volt battery, so it will be fully charged at all times. It is delivered with a two-year guarantee. Powered by a small group of photovoltaic cells atop the battery-pack, this unit can store enough power to operate for 21 days, even if the sun does not shine. The only requirement is the proper positioning of the solar cells when the unit is placed on its fencepost; they can't be shaded from the sky, and they should face south at the proper angle.

There is a solar-powered observation tower from which a United States Park Service forest ranger keeps watch over the forest in Lassen Volcanic National Park in California. There's a bunch of solar-heated swimming pools in San Diego, a solar-heated doghouse is being manufactured in Virginia Beach, a solar-powered automobile is being driven through the streets of Tel Aviv, solar telephones have been installed in Jordan, and a solar-cell music box for children that plays when it is put in the sunlight. That's being made by Solar International in Severna Park, Maryland—the same outfit that manufactures a solar-powered attic fan in its more serious moments.

Solar living is evolving at a pace not yet understood by the nation's political establishment in Washington, and indications of that progress go far beyond the manufacture of solar-powered toys or the invention of a solar-powered still. There is a village on the Papago Indian Reservation in southwestern Arizona some 30 miles north of the Mexican

The cost of power from the sun will go down as we develop better and cheaper ways of applying it to everyday needs. No foreign cartel can set the price of sun power; no one can embargo it.
President Jimmy Carter

border that now draws all of its electrical energy from a panel of photovoltaic solar cells installed in a cluster at the edge of the settlement of 100 Indians. If the unit had been installed when some of the first photovoltaic units were utilized by Skylab as its orbital energy source, it would have cost about $300 a watt for the electricity produced. When the Papago Reservation's plan was implemented in 1978, those costs had tumbled to $11 a watt. By 1979 they had been further reduced to $7, and late in that year, scientists at the Stanford Research Institute announced a breakthrough in the production process for pure silicone that they claimed would reduce the cost of photovoltaic cells by 90 percent.

"It's a lot like the electronic calculator or the transistor radio," says Bill Pearson, the director of the Indian Health Service in Arizona and a monitor of the Papago project. "It's quite possible that before too long you'll be able to buy a roll of photovoltaic cells like you now buy roofing paper, and have carpenters put it on."

Bill Pearson, for one, would not be put off by my vision of a home solar window that collects, converts, and regulates domestic solar energy. Such a window, after all, is one small step from the photovoltaic "roofing paper" that he foresees.

The problem, as I have suggested, is not technology, but politics; it's metaphysical, not physical. It has to do with the economic establishment, not the economics of energy. Some indicators of the mind-fix prevalent among the policymakers came almost immediately after the Stanford Research Institute announcement of a 90 percent drop in silicone production costs. There was remarkably uniform doubt at high levels of government and among utility spokesmen. "Important, if true," was the crux of their statements. The "if true" qualifier casts the pall of unreality over the entire story, helps persuade Americans who hear it that the silicone report is just another of those unreal claims about what the sun can do.

Which is why we must be so eternally beholden to the solar

tinkerers and inventors—the men and women who are introducing the sun to the citizens of their communities, the places where credibility starts. When a visitor steps into a solar-heated home and feels the warmth, when he is persuaded that there is indeed no furnace in the basement, when he is told that not a gallon of fuel oil or cubic foot of natural gas or a kilowatt of electricity is being purchased, then he

Is it so small a thing
To have enjoyed the sun,
To have lived light in the spring,
To have loved, to have thought,
to have done?
Matthew Arnold, *Empedocles on Etna*

Solar cells on Papago Indian reservation.

185

It takes a fully rounded individual to look at a problem and say: How can I make the answer simpler? Can I make it use less capital instead of more? Can I involve people in it instead of things? Can I make it a more human-scale technology—a technology with a human face?

From the film *The Power to Change*

becomes a believer. When the warmth is felt, when the sunlight is seen, then the conversion begins, then solar truths are suddenly credible.

There is not 1 person in 100,000 who can determine from the knowledge available to him who is right and who is wrong in the debate over the costs of producing silicone. If our government and industry spokespersons treat the claim with much scepticism, then so too will the public majority. On the other hand, if the 100,000 individuals who have visited the 40,000 existing U.S. solar homes are asked to evaluate the news from Stanford, they will be much inclined to believe it. They will because they have been exposed to solar realities; to them it will seem not only logical but inevitable that the solar progression continue. They would, quite likely, expect the cost of photovoltaic cells to continue to plummet, no matter how many subsidies are used to prop up the petroleum industry, or coal mines, or "synfuels."

Once an individual has been re-introduced to the sun as a presence rather than a statistic, solar living becomes credible, affordable, and most desirable. It is this unconquerable fact that is starting this nation on the soft energy, solar path. Every significant social change has worked from the people up, not from the politicians down. Living with the sun is no exception, and if ever there was an indication that our solar future is assured, it is the word from the United States Department of Energy that there are now tens of thousands of solar homes where, a short time ago, there were just 24.

"Last year," writes solar architect Malcolm Wells, "somewhere in Florida, on the leaves of a forgotten sugarcane plant, a bit of sunlight ended its eight-minute dash to earth. Somehow, the plant turned the sunlight into sugar. Somehow that sugar got into my sugar bowl and into my morning coffee. I sipped last year's sunshine at breakfast. Now it's in my blood, and it starts to feed these old architect muscles. It's dark now and I start for home on my bicycle. The muscled sunlight suddenly

becomes pedal-power, then chain-pull, wheel-spin, generator-whine, filament-heat, and, finally—from the headlamp—light again!

"Miracle!"

Yes, a miracle in many ways, but one that is commonplace for cyclists after dark. So commonplace that not many, besides Malcolm Wells, have ever made the solar connection. As more and more of them do, an entire population will enter the age of solar living, with, without, or in spite of, their government.

Goodness comes out of people who bask in the sun, as it does out of a sweet apple roasted before the fire.
Charles Dudley Warner, "Seventeenth Week," *My Summer in a Garden*

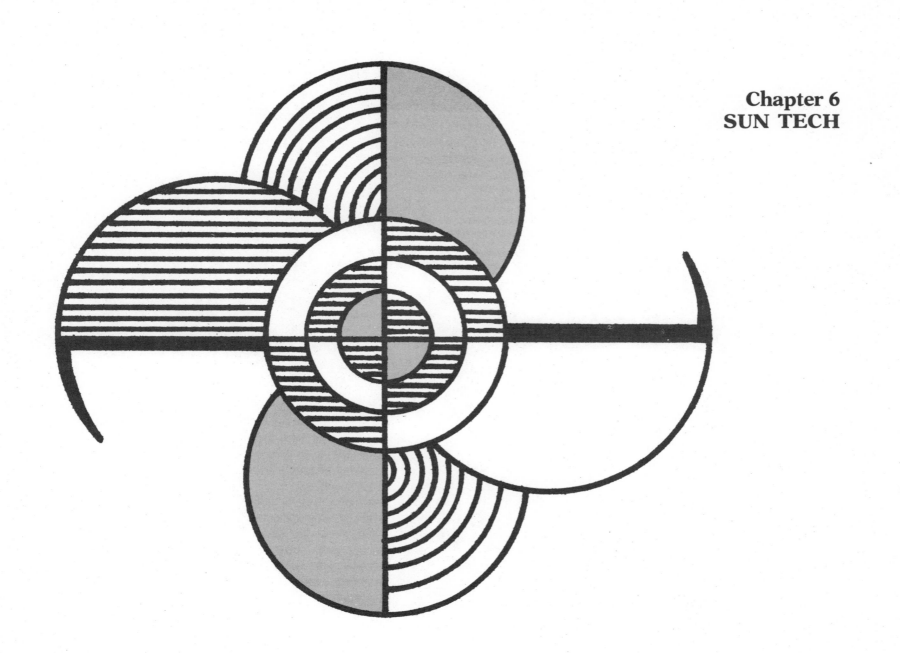

The sun never repents of the good he does,
Nor does he ever demand a recompense.
BENJAMIN FRANKLIN

6 *SUN TECH*

In the heart of downtown Milwaukee, and just beyond Boston's city limits, and at the rim of the Arizona desert are the concrete relics of a time the nation, and the world, shall not see again. Like pale ribbons cut by giant scissors, four-lane superhighways come to a stunning halt. Once, a decade or more ago, the cloverleafs, the divided highways, the elevated expressways, were carefully drawn by transportation engineers, developing with the limitless horizons of the petroleum age a road network designed to carry increasing numbers of passenger cars faster and faster.

Just as the designs reached an apex of intricacy and cost—more than a million dollars a mile—a recognition of limits seeped into the decision-making departments of state and national governments. Highway construction was halted, by sudden fiat. Men like Governor Francis Sargent of Massachusetts simply said, "Stop," and the bulldozers, graders, and ten-ton trucks ground to a halt, leaving behind them roads going nowhere, fingers of concrete on the green landscape, pointing toward a confused future—an era which would see automobiles going slower each year for the first time since their invention, an era which began with an end to 50 years of annual increases in the volume of gasoline sold. Long after alternate fuels are found, long after the solar automobile takes to the road—and leaves it—these concrete monuments to the energy transition of the seventies and eighties will remain. Like the statue of Ozymandias in Shelley's poem, fallen and abandoned in the desert, a relic of an

emperor who thought his reign immortal, the truncated superhighways of the sixties will remind many future generations of the decline of growth economics and the start of a Solar Age of Equilibrium.

Recognizing the omens several years ago, I wrote a brief essay on possible uses for some of the abandoned roads. Surfaced with asphalt, their right-of-ways cleared of all vegetation for 50 feet or more on each side, the roads seemed to me to be quite well adapted for use as solar receptors. At the time, I was unaware of the possibility that strips of photovoltaic cells could be laid along the highways like shingles on a roof. (That scheme, by the way, can still be used to supply energy to the electric cars of the future.) I was, however, aware of a growing shortage of fresh water, and suggested that the abandoned freeways of California could be used as immense evaporation beds. Seawater, I proposed, could be piped in and used to flood the roads at shallow depths. Evaporated by the sun, then condensed on special glass panels, the ocean water would be converted to pure salt and distilled fresh water—two commodities always in demand.

There was hardly a serious note in the essay; it was primarily a lighthearted look at the future and an attempt at enlightening some of my readers about the approaching twilight of cheap gasoline and relatively cheap cars.

Just recently, I was reminded of that column when I learned that the sunshine that falls each year on U.S. roads contains twice as much energy as do the fossil fuels currently being burned in all the countries of the world. Had I been more concerned with a factual instead of a fanciful approach to my essay, I might have grasped the true generosity of solar energy ten years sooner.

I was not alone, however. Indeed, it's my guess that the vast majority of Americans are quite unconvinced that solar energy can some-how fuel the great cities and the awesome industrial complex that are such pivotal presences in the United States. Oh, it may be possible (after

the legal underbrush has been cleared) to put a solar hot-water heater atop a surburban roof, but what about the steel mills of Pittsburgh, the copper smelteries of Montana, the lights of Los Angeles or New York? The sun has got to go big time.

I am reminded of an utterance of the fifties, given to posterity by Charles ("Engine Charley") Wilson, a past president of General Motors who began his public life as a member of the Eisenhower cabinet. "What's good for General Motors," he announced from Washington one morning, "is good for the country." Charles Wilson was much criticized for that remark. In my view, he shouldn't have been. He was, in his blunt way, articulating a fact which was true then, and is still true—although it's on the brink of being qualified by these changing times. What the former GM boss said, in so many words, underscored the truth about automobiles in America: they are the symbolic heart of this industrial nation. The annual gross income at General Motors is larger than the annual gross national product of most of the world's countries. When the assembly lines in Detroit slow down, so does the stock market; when GM production line workers are laid off, Congress and the president confess to the citizens that a recession has begun. And, as the eighties begin, General Motors is no longer merely a large automobile company; it quite dominates the industry. There are those who speculate on how many more years must pass before it becomes the only one in America.

In an industrial economy, growing and thriving industries are the base of economic stability. Most people understand this, which is why most people, when they can, buy General Motors stock, or United States Steel, or International Business Machines—the blue-ribbon investments that are likely to prove worthy as long as the nation holds together.

It is the almost unbelievable—to most Americans—instability of industry, the stock market, and the banks that has generated such widespread anxiety about the future. The reverse of Wilson's observation is also valid: what's unsettling to General Motors is likely to unsettle the rest of the country.

I dream of a solar future when our rhythms will be more natural, our pace less rushed. In my mind I see little solar cars running on tracks with an automated destination set. While riding in them at a slower speed than our present highway traffic, we'll be able to read, talk with each other, or spend some time in quiet meditation.
Careen Mayer, educator, Annapolis, Maryland

Solar panels overhead and on the prow charge batteries of this boat designed by solar energy expert John Hoke.

Yet an end to Charley Wilson's automobile industry was bound to come. The corporation which he helped build grew with the discovery of a single, nonrenewable resource—oil. As it has become more and more clear to more and more people that oil, and its refined products—gasoline, asphalt, heating fuel, and kerosene—are limited resources that

The sun glints on this grain drying project in Ohio which uses polyvinyl plastic solar collectors.

Hawaiian sugarcane field.

Solar collectors on a poultry house give 24-hour heat through using a water storage system.

Two methane digesters which convert the manure of 350 cows have been installed at the Monroe, Washington, State Honor Prison Farm. With a capacity of 100,000 gallons, the digesters are supplying this large installation with a good share of its heat and light.

In Iowa, all the heating, lighting, and pumping equipment that powers a chicken farm with a population of 165,000 birds is generated from the chickens' own manure.

Solar collectors heat the milking parlor and wash water at the USDA's Beltsville, Maryland, laboratory.

For the first time in history, you may be looking around the block for a place to park your car in the sun. This solar-powered car developed at Tel Aviv University runs on batteries charged by solar-cell panels on the roof and hood.

may last just another two or three generations, it has also become clear that the automobiles Charley Wilson built will have to be radically modified. With them, General Motors must also change, and when that fact is contemplated, nearly everyone begins to worry about what tomorrow may bring.

Most of the nervous Americans—and that's most of the population—cannot imagine General Motors as a solar-powered industry. If they could, they would not be nervous. It becomes clear, then, that yet another variation of the Wilson dictum is needed for these times. It should read something like this: if solar energy is good for General Motors, it's good for the country. Few of you could disagree with that. If you knew that the nation's largest industry was fueled by the sun, the sun would instantly acquire your full confidence as an energy source for all manner of needs. And if General Motors could use solar power to build a solar-powered car, the circle would be complete. There would be no doubt left in anyone's mind that we had indeed entered the Solar Age, and that all that talk back in the seventies about the limits of solar energy was as meaningful as the talk around the turn of the twentieth century when eminent experts announced that the occupants of any horseless carriage would be killed in their seats if the vehicle went faster than 30 miles per hour.

But what would it take to convert GM to solar, or to power Pittsburgh with the sun? How close are we to seeing these essential, large-scale applications of that diffuse energy that falls with such largess on our roads? And just how much energy are we talking about, anyway? What do we get from fossil fuels now that we will need from the sun in the future?

According to a report by the Edison Electric Institute, the United States will require about 150 quads of energy by the year 2000. A quad is 1 quadrillion (or 4,000 trillion) British thermal units (Btu.'s). Put another way, a quad is equal to the energy currently obtained from 172

million barrels of oil, 42 million tons of coal, or 293 billion kilowatt-hours of electricity.

Of that quad-use total, industry takes about 30 percent, when the electrical energy purchased by industrial consumers is added to the coal, gas, and oil burned directly at industrial sites. A total solar conversion would mean that the 50 quads of energy now supplied to factories and refineries would have to be generated by a combination of solar devices, solar systems, solar collectors, and related solar energy suppliers. That means that solar replacements would have to be found for about 86 billion barrels of oil.

These are difficult numbers to realize in terms that any of us can understand; 86 billion seems like an impossible, unrecognizable amount. Yet every aspect of the conversion of American industry would become credible, able to be understood and perceived, if just one General Motors assembly plant were able to get its work done with sun power. Detroit and its four-wheeled products have been an American metaphor for so long that their credibility is unimpeachable. Build one Chevy Citation on a solar assembly line, and you have convinced 200 million Americans that the solar idea can work.

But what is the likelihood that such projects can be realized? What are the solar appliances and mechanics—if any—that are able, or will be able, to get that first solar-powered car off that first solar-powered assembly line?

If history repeats—and there is no reason to believe it won't—those appliances and mechanics will grow from the smallest beginnings. Just as the internal-combustion engine and the horse-drawn buggy were first combined in Henry Ford's barn to become the horseless carriage, a solar-powered assembly line and a solar-energized city will not appear complete, ready to function. They will evolve, over the years—over the decades, perhaps—from work now being done.

The New Alchemists at Woods Hole, Massachusetts, make up

Solar cells on top of a road sign in Arizona provide power to rotate this sign during dust storm alerts.

197

Many Japanese industrialists are in the forefront in their country by supporting solar technology. They are planning and financing the building of over one million new solar homes and believe that by the mid-1980s they will be able to produce solar cells for less cost than fossil-fuel equivalents. One way of doing this may have already been discovered. Scientists in Japan have reported finding a highly efficient method of converting sunlight into electricity, using a component of spinach as an intermediary. Another recent development by the Japanese biologists at the University of Tokyo is the use of chlorophyll as the primary converter in experimental solar cells. Their tests indicate the process is 30 percent effective—much ahead of the 10 percent of more conventional, silicone-wafer photovoltaic cells.

one organization whose research has already made a difference. The "Alchies" might have once been regarded by governmental and professional communities with the same sort of scepticism reserved for young people with long hair and unorthodox clothing. The "hippie" image has dogged the New Alchemists since they first formed the group and set about their serious and scientific investigation into the physics and metaphysics of living productively in a postpetroleum age. Along with a good many other social attitudes which have undergone substantive change, the notion that groups like the New Alchemists are generally out of touch with reality is a notion that has lost its credibility.

The New Alchemists—and their counterparts around the nation and the globe—are being taken seriously. Their work—recorded in meticulous detail no matter what the project—has attracted interest in the capitals and universities of the world. And of the dozens of New Alchemists' efforts designed to instruct us in how to live in greater harmony with our natural world, none has been more provocative or more fully reported and discussed than the Ark, the bioshelter at Spry Point on Prince Edward Island.

Few of the millions of Americans who currently harbor qualms and anxieties about their postindustrial future can tell you the location of Spry Point on Prince Edward Island. Fewer still can tell you why this information about a remote spot in eastern Canada has any relevance. But as we cross the threshold of the eighties and enter a decade most often described as a tumultuous age of limits, shortages, and depletions, it may prove instructive if we learn more about the place on the Point—one of the few structures on the planet designed specifically to help us survive our transitional time.

I am not certain that John Todd quotes Scripture when he describes the Spry Point complex he helped inspire, but if he does, he most probably begins with: "Consider the lilies of the field. . . ."

Todd, 40, a founder of the New Alchemy Institute with his wife

In Kingman, Arizona, elderly people in a low-income housing project spent the winter of 1977/78 building solar greenhouses and enjoyed the first vegetables they had grown while their counterparts were worrying about the inflated costs of the same produce at the supermarkets. In conjunction with the same pilot program, the minimum-security prison in Kingman became the first institution of its kind to be fitted with a solar greenhouse, a solar hot-water system, and a tower for the wind generation of electricity. Built by the inmates, the prison's solar greenhouse is 60 feet long, 16 feet wide, built of cinder blocks, double-paned windows, and lumber salvaged from some old warehouses. Its builders and their cell mates are, literally, enjoying the fruits of their labor each time they share a meal which includes the fresh produce grown at their doorstep.

The vegetable growing area of the Ark with solar ponds.

199

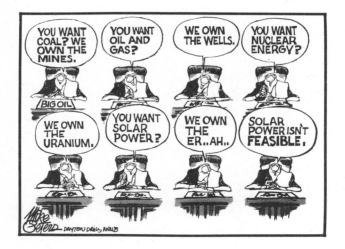

Nancy, puts the same thought in less biblical language when he talks about why plants are the model for the Ark—the bioshelter conceived in 1974, constructed the following year, dedicated and officially opened by Canadian Prime Minister Pierre Trudeau on September 21, 1976, and now a functioning organism of the Prince Edward Island community.

"If we create a way of living," says Todd, "a culture, in the image of the plant, if we emulate the biosphere, it would bring about a revolutionary change in the way people live on the earth. It would have an impact as great as the introduction of agriculture 10,000 years ago. "The Ark on Prince Edward Island is an early attempt to explore the landscape of this new solar synthesis."

The choice of Spry Point was one of the bolder decisions when the Ark's designers planned their demonstration of how human environments might be modeled after a plant's natural design. North of the fourty-sixth parallel, this easternmost section of Prince Edward Island is battered by winters as extreme as those in our northernmost border states. The location on the Gulf of Saint Lawrence is surrounded by pack ice during the winter and early-spring months, and the sun's seasonal azimuth is as low in December skies as it is in northernmost Maine and Minnesota.

The premise was: if the Ark works on Spry Point, it will work in most of the locations where most of the planet's people live. To "work" the Ark had to provide living space for a family, its own energy, its own food, heat, and light. It had, in other words, to be as totally self-sufficient as the vessel Noah constructed to ride out the Deluge.

That attempt was a joint effort of the New Alchemists and the Canadian government, whose Ministry of State for Urban Affairs responded with interest to pilot designs mailed to them early in the seventies by New Alchemists Hilde Mainge and Earle Barnhart. Those plans, in turn, were sent to Solsearch Architects on Prince Edward Island. Out of work done there by architects Ole Hammarlund and David

Bergmark came the unqiue, 7,000-square-foot building that combines living space with two greenhouses, fish-culture tanks, heat-storage tanks, solar collectors, laboratories, a barn, library, laundry, and a broom closet.

The Ark's interior climate that allows it to do just fine in Canadian winters that routinely include windy, below-zero sunless days and long nights is a result of an involved and sophisticated design. It is one of the most careful and scientific attempts to emulate a plant's natural vitalization of all that the sun has to give.

As you would understand it from a blueprint, the building is a long (just under 150 feet), two-story structure with a windowless side to the north and a vast expanse of glazing facing south, toward the sun. An earthen berm piled on the north side directs winter winds over the shed roof. On the southern exposure, vertical and 45-degree glazing cover almost the entire expanse, admitting light and heat, keeping out cold, and providing this particular bioshelter with the one essential for life on this planet—the solar presence.

It is that same solar presence that spawns the winds that turn the blades of the wind generator that supplies the Ark with some of its electrical energy. It is the solar presence which heats the 118 cubic yards of rocks stored in the basement under the barn, the 20,000 gallons of water in the 25 "solar algae-ponds" that line the greenhouse; and the solar presence that energizes the flat-plate collector that helps to heat the water in the 16,000-gallon storage tanks that warm the Ark's three-bedroom residential quarters. And it is, of course, that same solar presence that gave the growth to the trees that are cut and split into firewood for the wood stove in the Ark that supplies supplemental warmth on the days when the sun does not shine.

Like the natural plants which inspired it, the Ark is a sensitive, complex, and self-contained utilizer of the renewable solar resources that are its environment. Unlike the plants, it has not yet developed the wondrous synthesis which allows it to thrive with no apparent effort. To

keep the Ark functioning takes attention and labor; the structure is, however, as close to full natural syntheis as any yet devised.

The Ark is a documented example of how the best of technology and science can be utilized to develop systems for living in greater harmony with nature. The individuals who designed and built the Ark, and who record its evolution, are each trained and skilled at their work. Their records are dotted with scientific footnotes and references, their observations are exact, and whatever conclusions are stated are the result of the most classic Aristotelian adherence to rational thought.

The solar ponds that began, for example, as an idea of Todd's several years ago, have evolved to become a keystone of the Ark's function. Lining both sides of the large greenhouse that takes up most of the Ark's south side, the solar ponds are, in fact, cylindrical tanks made of highly translucent fiberglass. Their depth and diameter (both five feet) give them a capacity of 734 gallons—proportions sought to allow the maximum penetration of sunlight, and thus maximum energy storage.

But the ponds do more than trap heat for the plants growing in the greenhouse. They produce algae—phytoplankton—which in turn feed tilapia, an African freshwater fish that is an important protein source for the Ark's human family. And the cycle does not end there. It is, as Todd says, marvelously interconnected. The swimming of the tilapia keeps the water in the ponds moving, so the production of algae is enhanced. The fish, which consume the algae, grow almost five times faster than they would otherwise. As they grow, their own waste further enriches the water.

That water is used as a food for the dozens of species of plants and young trees that are grown in the greenhouse. Like a many-chambered nautilus, the intricacies of the Ark's diverse cycles follow from one to another to another until the total architecture of the environment's interior becomes as simple and yet as finely fashioned as a shell which has taken millennia to evolve.

Building a solar culture, not a solar industry, is the aim of the Solar Survival team in Harrisville, New Hampshire, one of the many small groups committed to exploring the potential of a solar lifestyle along with small scale technology. The group includes designers, carpenters and teachers who have come together to develop Solar Cones and Solar Pods (seed protectors and solar grow-frames), solar food dryers, and a novel passive solar house design. Working together on local solutions to local problems as this group does may be one hallmark of a coming solar age.

What in the world makes more sense than solar power? It will last as long as life on earth, until the sun swells out to become a Red Giant and consumes us in about four billion years. It arrives free from the generating station. It does not spill or leak. It does nothing more to the airspaces above our cities than brighten them.
Jacques-Yves Cousteau

Like the ponds which store energy and heat, the algae which convert sunlight to carbohydrates and oxygen, the fish which enhance the growth of algae, and, in turn, enrich the water that is used to feed the greenhouse plants, the plants and trees themselves also serve multiple functions. And, most importantly, together the total complex allows for a maintenance of natural interior temperatures that defy the Canadian winter cold. In December 1977, according to the institute's detailed records, outdoor temperatures reached well below 0°F. Inside the greenhouse, soil and air temperatures never dropped below 50°F., even after prolonged cloudy days and the region's long December nights.

Among the basic plants grown commercially in the large greenhouse—adjacent to the smaller "family garden" where the Ark's humans grow their staples—are lettuce, kale, spinach, chard, broccoli, parsley, bean, herb and flower crops.

A record of every creature found in the greenhouse, or introduced to it, has been kept and includes names familiar to every gardener: ants, bees, earthworms, earwigs, spider mites, toads, and wasps, to cite a few. The list of all the plants—vegetables, flowers, trees, and vines—that have been tested in the Ark runs to nearly 100.

Since the Ark at Spry Point began functioning, another has been built on 12 acres in Cape Cod. This one has no living quarters, but carries the horticultural lessons learned on Prince Edward Island to further degrees of sophistication. Still, the adventure has not ended for the Todds and their coworkers. They continue to be excited by their horizons; each day they see new visions for the future.

Just as the vast factories of Detroit and the world's largest industry grew from the seeds planted in Henry Ford's garage, so can the solar conversion of cities and assembly lines grow from the discoveries being made at Spry Point. When Henry began tinkering, not one of his neighbors believed he was founding a way of life. Few of them believed what he was doing would amount to anything except a contraption.

With the benefit of that hindsight, we might look at the Ark with, at the very least, some recognition of the principles it embodies, and some questions about how those principles can be "scaled up" so that they apply to urban and industrial settings.

But before you try to conceive your first sun-powered Ford, or a solar-energized Ford assembly line, you should look at the Ark in terms of what it does not do, rather than what it does. The Ark does not, for example, waste energy.

Conservation, then, as exemplified by the Ark, will be the first step toward the sun. More than three-fifths of all the energy used in the nation is consumed by industrial, urban, and commercial users. Of those, six industries have historically taken the lion's share. Those that refine and work with primary metals; the chemical industry; the food production network; the manufacture of stone, glass, and clay products; and the oil and coal refining processes (from which, paradoxically, come the energy burned by the other five) are the six giant, industrial energy users. However, much of the petroleum legacy bestowed to the planet millions of years ago is used inefficiently by these and other large commercial consumers.

Energy currently employed for various tasks within these manufacturing and refining operations is often of a higher quality than the job requires. Fitting the energy levels to the job, then, is one method the largest U.S. energy users could employ immediately. The Arks at Spry Point and on Cape Cod are excellent models. Not a single watt in those installations is applied to a function that does not somehow synthesize with the whole. Often, one unit of energy links with others to accomplish several tasks, one knit to the other across an entire spectrum of needs.

This kind of conservation is more sophisticated than householders can achieve by insulating their attics, or putting on storm doors, yet the net result is the same. Using energy sources capable of heating water to boiling and beyond when what's needed is water at 100° F. is a classic

Biogas plants have been a reliable source of power in China, India, and Africa for decades. The slurry left over is an excellent fertilizer, while the fuel can be used to run equipment, heat homes, cook and dry crops, heat farm buildings or power a generator to produce electricity.

Sunflowers as well as shrubs are being studied for their potential as fuel crops for the future. Sugarcane and cornstalk residues are first in line now for tapping the solar energy that comes to us by way of plant photosynthesis. Unlike direct solar energy which is otherwise unused, getting energy from plants involves us in the trade-offs between using biomass for food, animal feed, soil building, and fuel.

industrial example of energy waste inherited from a time when oil was three dollars a barrel and coal was so cheap that it didn't pay to mine it. Taking advantage of the technology that era helped make possible, a computerized system to supply low-heat fluids where they are needed in industry, and using only enough energy to achieve precise temperatures could save truly significant amounts.

The Ark at Spry Point is a web of monitors as well as a model unit of self-sufficiency. Very little change of any sort takes place behind the solar walls that is not measured, recorded, and, at some point, analyzed. But with the Arks as models of how to save megawatts, the American industrial complex can develop and install the same sorts of sensitive monitors, can develop the same delicate energy consciousness that allows every application to be analyzed in terms of fitting precisely the proper amount of energy to the job it's asked to do.

Such sensitivity is a key to the solar transition. With its available supplies of oil, coal, and gas, the United States has been a boorish energy user. Like an oafish heir to a fortune he did little or nothing to acquire, the nation has taken its energy bequests and spent them without regard, always believing there would be more, always sure that the way to any accomplishment was to burn enough fossil fuel, generate enough power. This way, minerals can be squeezed from the mountains that hold them, concrete can be piled 100 stories high, automobiles can travel at 100 miles per hour, and most of the office buildings in our cities can be constructed with windows that don't open. And when the fossil-fuel legacy that allowed such insensitive spending began to dwindle, when the heir reached into his energy purse and found it less than full, he turned to the only banker he thought could continue to finance more such squandering.

That banker deals in nuclear currency, and the reason nuclear power has such a grip on the industrial establishment is because it appears to offer wastrels the same extravagance guarantees that fossil

fuels once did. But we are learning, with each passing nuclear year, that the penalties of energy extravagance will be levied for centuries. There is an afterlife to the atom which cannot be avoided; the nuclear banker exacts a higher rate of interest than any we have imagined. The days of the energy boors are coming to an end; our fossil-fuel inheritance has been squandered.

The aftereffects of the binge are not, however, all bad. Just as some of the boldest and brashest tycoons of the early Industrial Age left museums, hospitals, and fine works of art in their indulgent wakes, so has our profligate energy era given us a highly developed technology which we can now use to learn to live with the sun. What's needed more than any technological fix is a shift in attitudes. And among the contributions to the solar future made by the New Alchemists, it may be that the attitudinal shifts their policies have set in motion will prove to be the most significant.

They did not approach the building of their prototypical model of self-sufficiency with the attitude that they would arrive at all the answers, the complete solution. Nor did they assume that every problem they encountered could be solved merely by throwing money at it, or feeding it more fuel. They realized, as the Todds have said many times, that they entered an unexplored world. They trod most gently indeed, and were prepared to make some missteps. They did not build a huge solar mirror to focus intense amounts of energy on a turbine which would generate steam to power an electrical generator. They were much more delicate with their approach to the sun; they worked as a man might work if he tried to emulate a spider spinning a web: with utmost patience and care, and with the understanding that each section of the evolving design would have to support another part of the fragile, yet functioning whole.

This is the first lesson cities and industries and nations will need to learn when they turn toward the sun for their energy. It may well

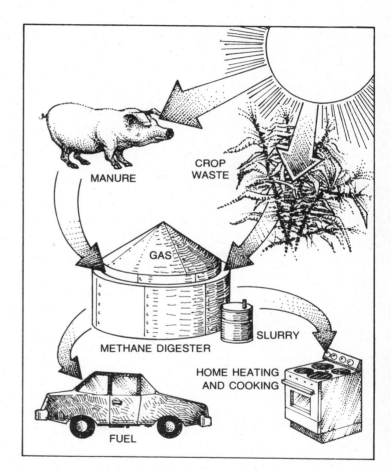

From the sun through plants and animals to power for us.

207

A newspaper in Florida has a wind system to power its lights, heat, and editing equipment; a Kansas farmer sells power from his windmill back to the power company when it's more than he needs; and large "farms" of wind machines are being developed out on the Pacific Northwest coast. While the initial cost is high ($8,000 to $12,000 in 1980), no industry as yet has figured out a way to put a meter on the wind. So once the windmill is in place, the source of its power, like the sun, is free.

be the most important lesson the Ark can teach us. The problem with solar energy is not in its gentle applications: sunlight pouring through a south-facing window can heat a room. We all know that now, although many people are still surprised by how warm the room can get. What we do not accept, what we, in fact, seem to resist, is the sun's capacity for energizing commuter transportation, television sets, automobiles, tenement furnaces, vacuum cleaners, hospitals, factories, jails, office buildings, and department stores—or whatever you visualize when someone asks you to imagine, asks you to think, about what's "big," what goes beyond the scope of your own shelter, or your own immediate and familiar community.

In these early days of our fossil-fuel hangover, it is difficult, if not impossible, for us to conceive of a system for capturing and distributing solar energy that can accomplish the "big jobs." On its surface, nothing in the Ark gives us a clue. Yet, if we look at the principles involved with some feeling for what's beneath the surface, we can find the models, the smallest beginnings, the images of the horseless carriage in the barn that gave Henry Ford and Charley Wilson their first insight into what they intuitively perceived to be the total automobile industry. They could not, and did not, foresee every varied form that industry would take, but they knew they were in touch with the future.

Such awareness, an understanding that the sun can indeed provide us with the energy and the life supports we need, is the enduring lesson of the Arks. It is a lesson yet to be learned by many of those who suggest that solar energy is wishful thinking, that we are somehow technically unable to use the storehouse in our sky. These attitudes, plus the efficient and determined work of the pro-petroleum lobbyists, have been the key to the national procrastination of solar-powered systems. Confidence in the sun is the key, and it is a demonstration of that confidence which the Arks so importantly provide.

Those translucent tanks that capture the sun to generate algae

growth that in turn feed the tilapia that in turn agitate and fertilize the water to generate more algae growth make a statement. It concerns the biomass: the solar application which converts energy from the sun to living plant organisms which, in turn, can be converted to other energy forms. These biological energy sources, which also include both organic wastes and fuel crops, could, by themselves, yield much of the commercial and urban world's current energy needs. These sources can provide liquid and gaseous fuels, as well as direct heat and electricity. Factories—a steel mill at Bethlehem, Pennsylvania, for example—could utilize biomass resources to supply both electricity and industrial-process steam. And, on the vast, energy-intensive farms which modern petroagriculture have created, enough straw is currently burned in the fields to supply every farm with the electricity it needs if that straw biomass were converted to energy instead.

According to an article in the February 1978 issue of the magazine, *People and Energy*, four major processes exist which can convert biomass to energy. Wood, garbage, and other wastes can be burned directly; most organic matter can be fermented to produce alcohol; sewage and manure can be converted to methane in digesters, which are tanks that provide a comfortable environment for anaerobic bacteria that convert sludge to gas; and organic matter can be pyrolyzed, a process which is analogous to methane digestion but happens at extremely high temperatures instead of with bacteria. Methane digestion and fermentation are highly efficient and produce no waste heat, but digestion is somewhat slow and unpredictable. Pyrolysis produces gas at high temperatures, oils at low temperatures, but is expensive, and it produces waste heat and materials.

The United States Department of Energy and the National Aeronautics and Space Administration tell us that as much as 30 quads of energy could be produced from biomass conversion. That's more than a quarter of the energy consumed in the nation, and much of the biomass

[*Continued on page 212*]

She's looking at a future source of oil . . . The shrub which produces this seedpod has a sap which contains hydrocarbons resembling the elements of petroleum.

The first American windmill was erected near Jamestown in Virginia.
The very first wind turbines were probably built in Persia as early as 200 B.C. Thirteenth-century crusaders brought the idea back to Europe. By the fifteenth century, Holland was using windmills to drain its marshes and pump its canals, literally powering itself by this indirect energy from the sun.

On July 11, 1979, the wind began spinning the blades of this turbine generator near Boone, North Carolina. Winds of 25 miles per hour can produce electricity for 500 average homes.

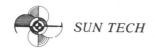 *SUN TECH*

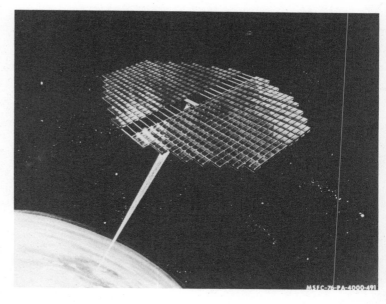

Orbiting solar collectors which would beam solar energy back from space are one high-technology possibility for a solar future. Advocates see space satellites providing baseline energy for industry and backups for other renewable resources. Opponents worry about effects of laser beams gone awry as well as the high cost for a space program.

The shores of Manhattan were lined with windmills in the early nineteenth century. They were used to power pumps and mills.

McDonald's does it all for you with a solar hot-water heating system in this Ontario, Canada, restaurant.

output is realized on a scale large enough to be useful to giant consumers like heavy industry, urban environments, and public transportation.

Large-scale applications of biomass energy may seem a long step from the delicate, green algae blooming in the sun traps behind the Ark's solar walls, but only the scale is altered. At the heart of the system is the sun's ability to transform organic matter; biomass conversion is, if you will, a daily emulation of the same sorts of changes that took place eons ago and transformed the earth's green carpet to what is now coal and oil. With algae, chicken and cow manure, straw, and just about anything else that is organically composed, the process can take place in a day, not a millennium.

If heat from the sun is supplied to a mass of organic material—and that material can be any plant leavings we now think of as waste—the process of fermentation begins, given proper containment and the addition of organic catalysts. Among the by-products of that fermentation are ethanol, methanol, and alcohol—clean-burning energy fuels that are the liquid residues of the photosynthesis and carbohydrate storage that began when the sun's rays first fell on a seedling plant, whether it was a blade of grass or a tall pine tree.

When the Ark was built, it simply wasn't respectable to suggest that plants might make an important contribution to our stockpiles of usable energy. But, since the early seventies, that has changed. Now, at least one member of the United States Senate Subcommittee on Energy Research and Development rates biological solar-energy conversion systems (another way of saying "plants" or "biomass") as one of two most promising areas of solar energy research. Already scientists have said that the amount of carbon dioxide fixed by plants each year represents ten times the fossil-fuel energy we currently consume annually. That's what the delicate, green lacework blooming through Prince Edward Island winters underscores; that's the connection that runs from Spry Point to Hawaii, the same sort of link that led from Henry Ford's garage to the automobile industry.

Biomass conversion is just one of the effective sorts of solar energy that can be used to energize large-scale America in the near future. Others have been researched and estimated by a number of scientific think tanks, among them the MITRE Corporation of McLean, Virginia, which does a good deal of energy studies for government agencies.

Wind energy conversion systems (previewed by the Ark's wind-driven, hydraulic power plant) will prove to be the most economical and widely used solar energy systems by 1985, according to the MITRE study. These systems are also expected to be the largest utility application of solar systems before the year 2020, primarily in the fuel-save mode for peak-load capacity without storage. Depending upon the government's incentive program, the electricity produced will range from 167 to 224 billion killowatt-hours per year by 2000.

In looking at total U.S. demand for electricity by year 2000, the MITRE report shows that there will be a need for 115 quads of primary energy with 6 quads replaced by solar technology. While 5.3 quads will be produced by solar hot-water and space heating systems, process-heat applications and wind energy conversion systems, the remaining 0.7 quad is expected to result from the use of solar thermal-electric utility systems and biomass for fuel.

The MITRE report notes that these are not insignificant contributions: "Twenty-two years ago the first commercial nuclear reactor was still on paper. Today, billions of dollars and 22 years later, nuclear energy still displaces less than half the 6 quads it is projected solar will displace after a like period of development."

If one is to judge from the work being done by groups like the New Alchemists and reports that continue to surface from varied sources around the nation, the development of solar systems with massive public and industrial applications is accelerating considerably faster than nuclear power ever has. In addition to these natural, organic systems, new man-made inventions are being tested. There are reports on solar energy

The Gump Glass Company in Denver, Colorado, has a rooftop solar collector that heats water. The water is used to warm some 7,200 square feet of office space that could get decidedly chilly in the Rocky Mountain winters if it were not for the new solar units. The company invested some $40,000 in the installation which now provides more than 90 percent of the building's heat.

In Alabama, a group of 100 low-income black farmers, members of the Southwest Alabama Farmer's Cooperative, are producing 40 gallons of organic fuel a day from contaminated grain useless for food production.

An Associated Press report (17 October 1979) that appeared in many national newspapers indicates in its own way just how much progress has been made in biomass conversion. "A bib-overalled farmer drove up to the Farmer's Cooperative Association in Cumberland, Iowa," the dispatch begins, "and walked over to a shiny fuel pump, pushed his straw hat back on his head and laughed.

"'How much is that gallon of corn likker anyway?' he asked.

"That was nearly two years ago, and the farmer was one of the first to try gasohol. Now, 625 Iowa service stations sell the fuel made from corn."

You must become an ignorant
    man again
And see the sun again with an
    ignorant eye
And see it clearly in the idea of
    it.
Wallace Stevens, "Notes Toward a Supreme Fiction"

collected by solar-powered satellites orbiting the earth and beaming the power by laser to continnental collectors. There are reports that the hydrogen of the oceans can be released by the application of concentrated solar energy, and reports that the flywheel, energized by the sun, will power our public transportation.

There is a pattern to all this solar experimentation; there is a pattern as clear and as sharp as the one etched by Henry Ford and his fellow inventors of the internal-combustion engine and the automobile. And just as those experimenters with a transitional energy source were able to move humanity from the preindustrial age to the Industrial Age, so too will the solar tinkerers now at work move us to a postpetroleum era. The question of whether solar industry can be applied to the large-scale, energy-intensive applications we have developed during the past 150 years is irrelevant—just as irrelevant as inquiring of Henry Ford whether his Model T would be able to be put out to pasture as horses were.

Of course the sun can energize our society. It will provide us with heat, light, transportation, shelter, commerce, manufacturing, recreation, poetry and the pursuits of pleasure. And it will do it on a scale that can meet the needs of the planet's 5 billion people. The problem is not the solar future, the problem is the fossil-fuel present. To break our petroleum habit, we are going to have to endure the trauma of change. Some of our leaders seem so anxious to avoid that change that they pursue policies designed to wait until the last drop of oil, and the last lump of coal, and the last cubic foot of natural gas are gone before we turn to the sun.

The reasons for anxiety about the solar transition are ill-conceived and counterproductive. We have only to look at our solar past, our solar reality, our solar heritage, and our solar myths to be able to sketch our solar future—our postpetroleum era that can be one of the most fulfilling of any of the ages of time.

Waterwheels, windmills, and solar buildings in a
vision of the future by artist Frank Bozzo.

215

**CHAPTER 7
SUN FUTURE**

They spoke I think of perils past,
They spoke I think of peace at last.
                    VACHEL LINDSAY

7 *SUN FUTURE*

If indeed the Solar Age is previewed in today's tenuous trends, how will our lives be changed by the transition? How will our star—that mysterious plasmatic mass of fusion secrets locked in caves of flame—affect our societal structures and ourselves? What metaphors are explained by the New Alchemists' solar ponds, what signals given by the rooftop solar hot-water heaters of suburbia? When the sun becomes the primary resource of our material lives, will its myths once again be restored? Will our natural renaissance change our view of technology? Will our well-being be enhanced or infringed as we move away from what the sun once locked underground and turn our eyes overhead?

It has eternally been the province of writers to describe the future. Many have stepped far from the present and entered a world of fantasy; others have taken minority movements and transposed them to the masses. Still others, journalists like myself, have followed trends and occurrences to their apparent, logical conclusions, and compressed time to peer into an age when those trends become mores and the occurrences become commonplace. That will be my process, overlaid, if you will, with a touch of fantasy—fantasy that is not, however, merely the child of my imagination, but the offspring of an understanding of the sun's fancy as well as its fact.

If logic alone is applied to present trends, any scenario for the foreseeable future must unfold in one of two patterns. Considering present cultural directions, alternatives for the future become distinctly limited.

We are, all of us on this globe, in rather general agreement that our banked resources—the fossil fuels and minerals the sun once locked within the earth's crust—are diminishing. The discovery of new, non-renewable resources is not a likely event. However, as a society which bases its economic system on incremental growth and is anxious about a future which does not promise more of the same, we are reluctant to abandon the resource base which makes the system possible.

For whatever reasons, human nature seems to prefer the misery it knows over an existence it can only imagine. It is the failure of contemporary leadership that none of our shakers and shapers has drawn reassuring sketches of tomorrow; instead, they talk only of the sacrifices that must be made if we are to survive today.

Such talk may seem to be inspirational, but it is also incendiary. A powerful nation, told that it is running out of oil while a relatively powerless nation still has a surplus, is likely to move most aggressively. Convinced at last that no more vast discoveries are yet to be made, those that have benefited most from the discoveries of the past will turn toward whatever resources are known, no matter whose they may be. Beginning to believe their own oratory about hard times ahead, leaders will be able to justify wars of survival if the darkest of the two available scenarios is followed.

But what of the other path, the nonnuclear, stable solar path, the road that can lead to a nonviolent future of relative peace and security? What will it take to get our leaders walking that road so we, in turn, can follow them, acting as if it is they, not us, who know this is the direction we should have taken all along? How do we, all of us, get moving on a path toward the sun? And what will life be like as we begin that journey?

This book is an attempt to begin to answer those questions. It is an effort to develop a faith in the solar path—a faith based on at least a rudimentary knowledge of the sun's scientific presence, the sun's place in our myths and religion, the sun's influence on the natural order of things,

on our personal well-being, on our shelters, and on the larger world of commerce around us.

This book is an effort to establish the sun's credibility by documenting the credibility it once had. Of all human generations, those of the last half century—a single drop in time's ocean—are the only ones which have lost faith in the sun. Persuaded by the absolute confidence in logic spawned by the Age of Enlightenment, awed by the discoveries of science, and propagandized by those who profit from the perpetuation of fossil fuels, the majority has been told over and over again that the solar path is an impractical one, to be followed only by dreamers and wishful thinkers.

And many have believed the propaganda. Displaying a disregard for history, an ignorance of the sun's place in their religions, and a lack of understanding of the sun's proven energies, some Americans are still persuaded that contemplation of the Solar Age is a foolish occupation. Given even the rudimentary documentation of the sun's cultural and scientific history that has been assembled on these few pages, one can only wonder at their ability to ignore the truth. Just as Galileo's telescope was denied its reality by a culture whose beliefs it upended, so the impressive credentials of the sun are refused by a twentieth-century culture that will not surrender its easy-energy past. Like a drowning sailor who clings to a drifting spar when he could have swum to safety, Americans often seem willfully determined to sink with fossil fuels rather than survive with their sun.

As these pages have hopefully made clear, however, there are still Galileos among us—men and women who will not be dissuaded, who will not be deterred from the pursuit of truth merely because some opinions are popular. Scientists who explore the sun tell us its energies are more vast than any we have yet discovered; physicians who research the sun's effects on ourselves explain that there are benefits still to be gained; naturalists who trace the sun's cycles through the earth's organisms are still wondering at its bountiful generosities; and the few individuals who have already made their break to the Solar Age have learned enough from their early adventures to prove that the solar world is not flat, that we will not fall off the edge, but that beyond our present, limited horizons, there is a new world of treasures and contentments, of natural orders and renewable resources able to be shared by all the earth's inhabitants.

This is the solar alternative. It does not promise a world in which all problems shall evaporate, but, on the other hand, neither does it lead as surely toward violence as the way marked by increasing scarcities of fossil fuels and increasing pressures toward fusion and fission.

The solar alternative, however, does include a revolution of sorts, and it is this aspect of the solar future which may most disturb those who would continue the present at any price. Perhaps if those who clutch so at the systems they know instead of moving to those which have not been properly explained could somehow travel in time until they reach the Solar Age, they would come away reassured that the sun is no cause for anxiety.

The Solar Age will be different, there can be no doubt about it. This is why we need leaders who can outline those differences as reassurances, who can promise something besides tough times ahead. In my view, based on an extension of current trends now being explored by solar pioneers, most of the differences are for the best. Coupled with the technologies perfected during the past decades of easy energy, the postindustrial era which the Solar Age will bring can combine the best of both worlds. Or, put another way, living in harmony with the sun's rhythms can do a great deal to eliminate the discords of pursuing incremental growth. We shall be the better for it.

What are the changes we can anticipate as we move toward a solar culture? There are many, they are large and small. They will affect our inner beings and our external lives. We will become, over the solar

centuries, a different entity than we now are in these waning days of the age of nonrenewables—an age which has developed, flowered, and declined ever since Stone Age man discovered molten copper running from the embers of his fire.

Since then, the finite resources have tended to become increasingly finite in the sense that we have come to realize their limited nature and thus, their increasing value. Once the planet was circumnavigated and surveyed, an awareness of the importance of real estate became a primary motivation. And as more and more resources became the property of fewer and fewer individuals, the process of centralization accelerated. Now, as the Industrial Age decelerates and nonrenewables reach their highest perceived values, that centralization has become intense. A handful of oil companies and a handful of nation states now control most of the energy consumed by the entire world. Governments have followed the patterns set by commerce; communities have watched their independence erode. We are a nation propelled by the incremental growth of our centralized institutions. Charley Wilson knew that the principle to be true when he said, "What's good for General Motors is good for the country."

Energy in the hands of the oil companies and the nuclear power combines has become highly centralized, tightly focused, and increasingly costly as it becomes increasingly limited. The way we live is shaped by the energy we consume to keep us alive. Thus we are rattled by economic uncertainty as we learn our energy supplies grow increasingly uncertain. Thus we build more and more centralized institutions as we recognize our dependence on centralized resource distribution. The patterns of our culture are shaped by our cultural consumption.

Which is why our step toward the sun will take us in quite an opposite direction. The sun is diffuse; the sun is decentralized; the sun is orderly, predictable, steady, as unchanging in the patterns of its 10 billion years as any presence we have yet discovered. A move toward the sun as an energy source, as a primary sustainer of life, will revise the cultural hallmarks of the Industrial Age.

We will enter a time of equilibrium, an era of cycles as rhythmic as the seasons. Instead of the rising line of incremental growth that angles across any graph of human events—economic and biologic— our existence will be charted by a series of gentle curves that meet each other to make a circle, just as the sun, over the age, tracks the same orbits it followed eons ago. We will learn, by living with the sun, to accept the steady-state economy as a harmonious system, just as much a part of the natural order as the changing seasons and the passing days.

This does not mean, as some say, that our lives will lose animation, that there will be some sort of governor placed on the human spirit. Just as the days of May and September reach extravagant degrees of excitement, stirring, and beauty, so can lives lived attuned with natural presences. The primary criticism of the steady-state economy is that it will somehow smother initiative. It won't, but initiative will follow new directions. We are creative creatures. The denial of nonrenewable resources for the creation of industrial empires does not mean there are not other forms of creativity. Finding ways to allow us all to live together more productively, to continue the pursuit of the arts, the education of the young, the further evolution of good health, and the invention and design of shelters for the Solar Age will demand as much imagination and ingenuity as any age of the past. What will change is the results: instead of individual material acquisition, the community will be the primary beneficiary.

Being a diffuse, universal resource, the sun removes the incentives for one community to turn on another in a battle over particular, centralized, and scarce energy resources. Transposed to the level of day-to-day living, this basic solar truth means there will be broad-based changes in the structures we live in as well as the community systems we live by.

Our homes are the microcosms of our larger social structures. A majority of us now live in places built in the 1950s—a time when the notions of limits were hardly heard, much less discussed. Thus our shelters were interconnected, not only regionally, but globally. They were nourished by centralized systems fueled by readily available, low-cost, nonrenewable energies. If you lived in Maine, your citrus fruits came from across the nation, your home heating oil from across the oceans, and the products in your department store were shipped from around the world. Technology had taken over from nature, and the dwellings reflected this fact of the Industrial Age.

"Central heating" was the foremost staple of the shelter standards; homes were built to defy and conquer nature rather than existing in harmony with it. And the amount of energy those homes consume in the process of ignoring nature is extravagant. But that didn't matter in the 1950s. Oil was less than 20 cents a gallon; in relation to the natural forces which had been consumed over the eons to manufacture that oil, we were, all of us, being given a free ride. We could conquer winter with a turn of the thermostat, we could conquer distance with a flight aboard a jet plane, we could make anything work by supplying enough power, and our communities could become places to live rather than frameworks for living because the energy available allowed us to leap time and distance, borders and boundaries.

By the 1990s—a time when even the most stubborn will have to accept the notion of limits and the end of easy energy—the majority of the homes in the nation will have been built in the 1970s. Of the millions, as we have learned, less than 100,000 homes will have been built for the solar alternative—unless a great deal changes in the interim.

If it does, if the building of solar homes is accelerated by people like the New Alchemists and by responsible leaders in government and by the increasing postindustrial momentum, then the trauma of change will not be so severe. If, however, the importance of the solar opportunity

[*Continued on page 228*]

AN OPTIMAL SOLAR STRATEGY

Residential and Commercial

End-Use Energy Form	Application	Percentage of Overall Energy Use	Appropriate Energy Supply Technology
Low-temperature thermal energy (< 100°C)	*Space heating, water heating, air conditioning*	*20–25%*	*Passive and active solar systems, district heating systems*
Intermediate-temperature thermal energy (100–300°C)	*Cooking and drying*	*~5%*	*Active solar heating with concentrating solar collectors*
Hydrogen	*. . .*	*. . .*	*Solar thermal, thermochemical, or electrolytic generation*
Methane	*. . .*	*. . .*	*Biomass*
Electricity	*Lighting, appliances, refrigeration*	*~10%*	*Photovoltaic, wind, solar thermal, total energy systems*
	Subtotal	*35%*	

Industrial

End-Use Energy Form	Application	Percentage of Overall Energy Use	Appropriate Energy Supply Technology
Intermediate-temperature energy (< 300°C)	*Industrial and agricultural process heat & steam*	*~ 7.5%*	*Active solar heating with flat plate collectors, and tracking solar concentrator*
High-temperature thermal energy (<300°C)	*Industrial process heat & steam*	*~17.5%*	*Tracking, concentrating solar collector systems*
Hydrogen	*. . .*	*. . .*	*Solar thermal, thermochemical, or electrolytic generation*

Industrial (Continued)

End-Use Energy Form	Application	Percentage of Overall Energy Use	Appropriate Energy Supply Technology
Electricity	Cogeneration; electric drive, electrolytic, & electrochemical processes	~10.0%	Solar thermal, photovoltaic, cogeneration, wind systems
Feedstocks	Supply carbon sources to chemical industries	~ 5.0%	Biomass residues and wastes or plantations
	Subtotal	~40.0%	

Transportation

End-Use Energy Form	Application	Percentage of Overall Energy Use	Appropriate Energy Supply Technology
Electricity	Electric vehicles, electric rail	10–20%	Photovoltaic, wind, & solar thermal-electric
Hydrogen	Aircraft fuel, land & water transportation vehicles	. . .	Solar thermal, thermochemical, or electrolytic generation
Liquid fuels–methanol, ethanol, gasoline	Long-distance land & water transportation vehicles	5–15%	Biomass residues and wastes or plantations
	Subtotal	~25%	
	Total	100%	

Steven Nadis, "An Optimal Solar Strategy," *Environment,* November 1979

continues to be minimized, then there will be a wild and painful scramble for solar homes when fossil fuels are priced off the market. Whichever path is followed, however, sometime after the turn of the next century, a majority of the temperate-zone population will be living in solar-oriented shelters.

In every temperate-zone research center on the planet, someone is at work on some aspect of solar conversion. In spite of haggling over the precise numbers of barrels of oil left to be pumped, in spite of talk that there is still hidden oil waiting to be extracted from seawater, in spite of the claims about shale reserves, tar pits, and oil manufactured from coal, those who think logically know that these petro-promises are only a bridge from fossil fuels to solar fuels. What is waiting to be mined, or pumped, or blasted from the earth is finite and nonrenewable. And it is to that basic fact the need for the sun is anchored. We are talking about a switch from nonrenewables to renewables; that is the simple, unarguable truth. The only debate is over the length of time we have to make the switch.

Scientists are aware of logical truths; such is the nature of their work, and has been since Aristotle. They know that if today's oil reserves were doubled or tripled, the time given us to burn that energy will be a mere tick of the cosmic clock. There will be just two choices left to us: emulate the sun here on earth and try to master fusion; or take what the sun is already sending us and convert it, via one system or another, to energy that will move us from one place to another, to energy that will heat and light our structures, and power the wheels of industry and commerce.

We have learned already that the business of nuclear fission and fusion can be expensive and risky. We have also learned that it tends to draw us to an even more centralized system. With most of our energy coming from just a few sources, our other institutions will conform to the same pattern. Our cities will grow even larger, our government more

top-heavy, our institutions more overweight, and the individual and his small communities less and less influential.

Such matters, however, are of little or no concern to the solar researchers. Ways of living with the sun will evolve unscientifically after the researchers make their solar discoveries, just as ways of living with fossil fuels evolved from beginnings as unlikely as Henry Ford's barn. Surely the automobile as the dominant mechanism in American economic and social life was one of the least-likely future visions John D. Rockefeller could have had when he began selling crude oil from his first wells in Pennsylvania. His only market then for the foul-smelling, black goo was as a substitute for whale oil; then becoming increasingly expensive as the Starbucks and Ahabs of New Bedford kept killing the sea herds that a previous generation of Yankee whalers had said were too numerous to ever become scarce.

We must, then, think of the solar scientists as the Henry Fords of today. We should take it for granted that from their work will come the conversion systems that will allow us to live gently with and from our sun. Talk of which mechanisms, which systems, will do the job is likely to be as unproductive as the debate at the state of the twentieth century over which kind of power—steam, internal combustion, or electric—would move the bulk of the horseless carriages that would surely evolve from the prototypical engines then being designed. There was, to be sure, much debate over the choice, but in terms of the nonmechanical effects the car had on every single American, that debate proved irrelevant.

What we should be doing is contemplating the social and environmental changes that will be wrought by our solar future, instead of placing figurative bets on whether the photovoltaic cell or biomass conversion will be the system we eventually utilize to capture the sun. With every basic technological shift have come social upheavals of lasting magnitude. But in spite of the printing press, mass production, the repeating rifle, the internal-combustion engine, and each of the other mechanisms which change the way we live and the way we view our lives, we still talk of solar energy in terms of its net costs, its payback time, and the "engines" which will produce it.

But the exploration of the socioeconomic designs that we can expect to live with under the sun is a research area even such advanced solar explorers as the New Alchemists are only beginning to sketch. Nevertheless, some general dimensions are beginning to emerge from the New Alchemists' work and the work of their several counterparts in this nation and others.

Basic to any description of the solar future is the matter of growth versus equilibrium. As I mentioned earlier, this transition—or even a mere discussion of it—seems so shattering to a United States nutured on six generations of ever-increasing exponential growth that any serious talk of an equilibrium economics, a steady-state system, is considered political suicide by those in office, or, even more, those who must run for office. The matter is not a campaign issue; it is not often mentioned in federal or state legislatures.

And because it is not mentioned, because it is, no matter what else it may be, quite definitely a change, the equilibrium system has come to be construed as negative—bad because it's different. One of the first steps that could be taken to boost public confidence in the Solar Age of the future would be to explain the equilibrium economy fully enough to ease the doubts of all those who think that the present system of incremental growth is the only way humans have ever devised for allowing each to live with himself and each other.

The facts are: one, the increments of growth by which the United States economy is measured have grown smaller and smaller as we approach the end utilization of the planet's nonrenewable resources; and, two, the ideas that next year should always be bigger than the one just ended, that we shall have a lifetime of raises, that tomorrow will be even better than today are, in their totality, ideas that do not harmonize

with natural law. In addition, they are concepts which accept the thesis that no matter how happy or content we may feel today, there is always the real likelihood that if we "live right" we'll feel even better tomorrow. The reverse, in reality, has often proven true. Raised expectations about the rewards of incremental growth have proven false so often that instead of being content with life as it is, many Americans are quite discontented, always waiting (querulously to be sure) for the tomorrow that will embody the dreams they have been promised will come true. What's happened is a kind of nationwide dissatisfaction, an endemic cynicism which must send many Americans to their graves wondering what went wrong, asking themselves where they missed the boat that they were told would carry everyone to "better living through chemistry."

The pattern of socioeconomic expectations parallels the graph that documents the gross national product, or the total number of oil barrels emptied each year. That pattern follows an inclined plane that moves upward, from right to left, in reasonably predictable increments—at least it did until the term "energy shortage" became part of the national lexicon. As we all know, there is no shortage of energy—the sun sees to that—but there has been, over the past several years, a shortage of consciousness that has altered our economics. By refusing to admit to the inevitability of a depletion of nonrenewable resources, the leaders of commerce and government have made certain that we will stay coupled to a growth system even though the resources which make it possible are no longer available.

If you would like to paint your own portrait of a solar future for you, your children, and their children, then the best place to begin is by understanding in your own mind just how and why that future would be different from the present. And don't make judgments about which is better, or worse. That's where we always get into trouble. Look at what can be, what will be, and the decisions about comparisons will be made automatically.

If the sun brings us closer to an equilibrium economy, as it will, how will our lives be altered? No matter what your life is like now, you can answer that question for yourself by comparing an inclined line to a circle, by comparing a growth system to an equilibrium system, by comparing an exploitation of nature to harmonious natural relationships, and by comparing a high-energy system of mass production, mass consumption, and mass waste with appropriate energy systems, community production, careful consumption, and almost no waste.

Among the more significant changes brought by the solar society, none may prove more welcome to the human family than an easing of international tensions—tensions exacerbated by the shortage syndrome of the last days of nonrenewable economics. Mexico, Venezuela, and the Middle East nations, for example, countries which had little before they found oil, would be less vulnerable to the exploitation and violence that will follow when oil-short superpowers realize their supremacy hinges on still more oil. Because the sun is diffuse, because its rays are distributed in equal capacities throughout the latitudes, scarce resources would no longer be so crucial to a nation's existence. The solar race would not be an arms race, but a competition to see which nations could conceive and construct the most efficient catchers and keepers of the sun's energies. As each country develops the systems for its own energy supplies, each will also develop the true independence which must precede global peace.

Which is not to imply that a solar society will disarm, that a nonviolent utopia will unfold merely because the quest for resources wanes. Human nature is not that easily modified. But living with the sun will increase the chances for peace; living with a diminishing-fossil-fuel economy will increase the chances for war. And, when solar systems do become perfected, the emphasis on nuclear power will also shrink, thus denying the military materials for fission and fusion that are now by-products of atomic generation.

The sun's diffusion also insures the survival of the nation's small communities. Indeed, until solar technology reaches its zenith and is able to concentrate the amounts of energy cities must have, the trend toward large urban centers will be reversed. With 90 percent of the American population currently living on 10 percent of the land, a renewal of the countryside will be a welcome ramification of a burgeoning solar society. High-energy urban environments have proven negative on balance; cities like New York, Chicago, Detroit, and Los Angeles have demonstrated that they create random violence, anxiety, and depression among the majority of their low- and middle-income populations.

Living by the sun will do more than supply an energy-hungry society with the power it needs to maintain an accustomed standard of security; living by the sun will feed spirits starved for some contact with the natural order of things. And as that nourishment begins, the anxieties which the industrial emphasis has generated will diminish; all peoples will find it easier to live with themselves and each other.

It is unfortunate, to say the least, that our leaders are so attached to the whipsaw of incremental growth that they are unable to visualize a brighter future under the sun. Instead of recognizing how much better a solar society can be, they tell us, with dismal repetition, of the "hard times" ahead, of the self-denial that we shall each have to endure. Such forecasts not only add to the already painful degrees of anxiety that trouble the populace, but they are demonstrably inaccurate. They are, at face value, quite wrong.

A solar future will be a future that helps to close the gap between the haves and the have-nots—gaps widening each day now as the cost of fossil-fuel energy leaps from month to month. A solar future will help revitalize small communities and their agricultural heritage. It will re-introduce an entire generation to the peace of natural rhythms, to a comprehension of cycles that allows for a calmer sort of individual fulfillment and a more natural, mature attitude toward death. The notion

of immortality as a privilege has been spawned by the technological age; an understanding of life's cycles will come as we turn toward the sun.

Nations will find a greater basis for coexistence as they share a sun that shines for all. Human health will improve as the skies, the land, and the water are less polluted by energy-intensive industry. The architecture of our shelters will allow us to be more conscious of the wonders of the world around us. Employment will increase as the solar conversion grows and its labor-intensive applications are perceived. Low-cost, low-energy mass transport will be restored; highway deaths will decrease. A service-oriented economy will replace a production-based system; high-risk jobs which use human beings as an accessory to production-line machines will give way to community-based employment that pivots on human relationships.

For all but a tiny percentage of the populace—those who now profit excessively from a world of waning nonrenewables—a solar future will be a brighter future. It is only the transition from the Industrial Age to a postindustrial society that is in doubt.

The sooner we turn toward the sun, the greater the potential for a peaceful transition. A solar future cannot be dark. By its very nature, it glows with promise. Humans in every corner of the planet have known that since Faventinus and before. Architecture, religion, logic, science, myth, and a sense of comprehension of the natural order each have the sun as their source.

The sun's history, the wonderful solar tapestry that has been woven through all the ages, is proof that we must set our recent past in its proper perspective, that we must use the best of the technologies those decades have spawned to restore our solar balance. It is time. If the decline of fossil fuels has told us nothing else, it has told us that. The only wonder is how slow we are to get the solar message. After all these centuries, you would think we would have learned to live more gently with our star.

Bibliography

SUN FACT

Eddy, John. *A New Sun, The Solar Results from Skylab*. Washington, D.C.: National Aeronautics and Space Administration, 1979.

   A look at sun science and history with a focus on solar discoveries and photography from Skylab.

Moore, Patrick. *The Sun*. New York: W. W. Norton, 1968.

   Designed for the intelligent amateur astronomer, this is a description of the sun and its effects on earth, including notes on the great solar astronomers.

Sagan, Carl. *The Dragons of Eden*. New York: Random House, 1977.

   An in-depth look at theories on the development of human intelligence.

von Braun, Wernher, and Ordway, Frederick. *New Worlds: Discoveries from Our Solar System*. New York: Doubleday, 1979.

   Current theory on the birth of the solar system and the lives of its family of planets.

Von Ditfurth, Hoimar. *Children of the Universe*. New York: Atheneum, 1976.

   Surveying what we know of the place of our earth in the universe, this book suggests that understanding the physical unity of the cosmos can help us restore our sense of wholeness.

SUN FANCY

American Museum of Natural History. *Anthropological Papers: Notes on the Sun Dance of the Sarsi, Cree, Dakota*. New York: American Museum of Natural History, 1919.

   The result of visits to the tribes, observations of sun dances, and collections of stories about sun dance events and history.

BIBLIOGRAPHY

Campbell, Joseph. *The Masks of God: Oriental Mythology*. New York: Viking Press, 1962. *The Masks of God: Occidental Mythology*. New York: Viking Press, 1964.
  Explanations of myth, art, worship, and literature—East and West.

Hawkes, Jacquetta. *Man and the Sun*. New York: Random House, 1962.
  From the rainbow light of prisms to Aztec sun worship—a survey of sun science and beliefs.

Helfman, Elisabeth. *Signs and Symbols of the Sun*. New York: Seabury Press, 1974.
  A children's book on sun myth, art, and history which can be enjoyed by adults too.

Larsen, Stephen. *The Shaman's Doorway*. New York: Harper & Row, 1976.
  Reflections on the mythic imagination and the place of myth in our lives.

Olcott, William Tyler. *Sun Lore of All Ages*. New York: G. P. Putnam's Sons, 1914.
  An extensive collection of sun myths, descriptions of sun religions, and compilation of sun lore.

SUN NATURAL

Calder, Nigel. *The Weather Machine*. New York: Viking Press, 1975.
  A source book on the causes and effects of weather which deals with major global patterns as well as local phenomena.

Cook, J. Gordon. *We Live by the Sun*. New York: Dial Press, 1957.
  Description of light and color in the world of vision and natural light rhythms.

Graham, F. Lanier, ed. *The Rainbow Book*. New York: Random House, 1979.
  Art, physics, and metaphysics of the rainbow.

Moore, Shirley. *Biological Clocks and Patterns*. New York: Criterion Books, 1967.
  Easily readable and entertaining book on plant, animal, and human rhythms.

Ward, Ritchie. *The Living Clocks*. New York: Alfred Knopf, 1971.
  Biological clocks, including light-related ones, in plants, animals, and us.

SUN PERSONAL

Giese, Arthur. *Living with Our Sun's Ultraviolet Rays*. New York: Plenum Press, 1976.
  Sunlight and health with a study of how ultraviolet rays affect skin cells and sometimes cause cancer.

Luce, Gay Gaer. *Biological Rhythms in Psychiatry and Medicine.* Chevy Chase, Md.: National Institute of Mental Health, Pub. #2088, 1970.
  Detailed scientific survey of cyclic rhythms in human life, including sleep, nutrition, and stress rhythms.

Mander, Jerry. *Four Arguments for the Elimination of TV.* New York: William Morrow, 1977.
  Along with considering the effects of TV, Mander looks at how natural light affects our health.

Ott, John. *Health & Light.* Old Greenwich, Conn.: Devin-Adair, 1973.
  The story of Ott's discoveries of light–health connections.

SUN LIVING

Anderson, Bruce, and Riordan, Michael. *The Solar Home Book.* Harrisville, N.H.: Cheshire Books, 1976.
  Energy conservation, passive and active solar design, both large and small scale, are clearly described.

Bainbridge, David; Corbett, Judy; and Hofacre, John. *Village Homes' Solar House Designs.* Emmaus, Pa.: Rodale Press, 1979.
  A realistic look at a working solar subdivision showing energy-efficient, solar-heated and -cooled housing designed for middle-income families.

Butti, Ken, and Perlin, John. *The Golden Thread.* Palo Alto, Calif.: Cheshire Books, 1980.
  Two thousand years of solar energy around the world from Greek to Indian civilizations, and how it's developing in ours.

Center for Renewable Resources. *Solar Energy Education Bibliography.* Washington, D.C.: Center for Renewable Resources, 1979.
  A list of solar, wind, water, and biomass books for elementary, secondary, and adult reading levels.

McCullagh, James C. *The Solar Greenhouse Book.* Emmaus, Pa.: Rodale Press, 1978.
  How to construct a solar greenhouse in any climate and for a variety of needs.

Skurka, Norma, and Naar, Jon. *Design for a Limited Planet.* New York: Random House, 1976.
  Pictures and short descriptions of alternate energy homes featuring interviews with the people who live in them to uncover the spirit of solar living.

Vories, Rebecca, and the staff of Solar Energy Research Institute. *Reaching Up, Reaching Out: A Guide to Organizing Local Solar Events*. Washington, D.C.: U.S. Government Printing Office, #061-000-00345-2.

How to build a local solar organization, raise money, and spread the word about solar energy.

Wade, Alex. *Design and Construction Handbook for Energy-Saving Houses*. Emmaus, Pa.: Rodale Press, 1980.

A handbook for designing, financing, and building a custom-made, energy-efficient, and reasonably priced house.

Wade, Alex, and Ewenstein, Neal. *30 Energy-Efficient Houses . . . You Can Build*. Emmaus, Pa.: Rodale Press, 1977.

A guide to designing and building houses which are space-saving and energy-efficient, complete with floor plans, drawings, and specifications.

Wells, Malcolm, and Spetgang, Irwin. *How to Buy Solar Heating . . . without getting burnt!* Emmaus, Pa.: Rodale Press, 1978.

A consumer's guide to choosing, financing, and installing solar home heating equipment.

Wolf, Ray, ed. *Solar Growing Frame*. Emmaus, Pa.: Rodale Press, 1980.

A detailed guide to building and maintaining a solar growing frame which includes plans, building instructions, materials list, gardening instructions, and seed packets.

SUN FUTURE

Commoner, Barry. *The Politics of Energy*. New York: Alfred Knopf, 1979.

How a national government could ease the transition to a solar economy by making big oil firms into public utilities and moving toward economic democracy.

Hayes, Denis. *Rays of Hope: The Transition to a Post-Petroleum World*. New York: W. W. Norton, 1977.

The arguments for turning to the sun, including a look at how the solar transition will shape our lives.

Lovins, Amory. *Soft Energy Paths*. New York: Harper & Row, 1977.

Why we should choose renewable energy sources over nuclear power and a path through the choices to be made for that goal.

Art Credits

page
   v   Illustration by Jerry O'Brien

   9   Courtesy of the New York Public Library Picture Collection

10, 11 Courtesy of the American Museum of Natural History

12   Smithsonian Institution Photo No. 56,122

13   *Left:* Courtesy of the National Aeronautics and Space Administration
Right: From *Music of the Spheres* by Guy Murchie, published by Houghton Mifflin Company. Copyright ©1961 by Guy Murchie. Reprinted by permission of the author and the publisher; British Commonwealth Publishers—Rider Books.

14   *Left:* Courtesy of the New York Public Library Picture Collection
Right: Smithsonian Institution Photo No. 46,834-D

16   Courtesy of the American Museum of Natural History

17   Smithsonian Institution Photo No. 66,029

18   Courtesy of the New York Public Library Picture Collection

20   Illustration by Mark Schultz

22   Courtesy of the Library of Decorative Arts, Paris

23   By permission of the Houghton Library, Harvard University

24   Mount Wilson and Las Campanas Observatories (Mount Wilson Observatory photographs)

25, 26 Courtesy of the National Aeronautics and Space Administration

27   Courtesy of the New York Public Library Picture Collection

28   *Center:* Courtesy of the National Aeronautics and Space Administration

page
Lower right: From "Interplanetary Particles and Fields" by James A. Van Allen. Copyright © 1975 by Scientific American, Inc. All rights reserved.

29, 30 Illustrations by Mark Schultz

31   Mount Wilson and Las Campanas Observatories (Mount Wilson Observatory photograph)

33   *Left:* Smithsonian Institution Photo No. 48,244
Right: Courtesy of the National Aeronautics and Space Administration

34   Courtesy of the Brookhaven National Laboratory

35   © S.P.A.D.E.M., Paris/V.A.G.A., New York, 1980; Courtesy of Musée Marmottan, Paris

37   *The Voyage of Life: Old Age* (2553)
Thomas Cole
National Gallery of Art, Washington
Ailsa Mellon Bruce Fund

38   Illustration by Tina Sullivan; from *Home Energy Digest and Wood Burning Quarterly*, Winter 1978, vol. 3, no. 3.

39   *Left:* The Phillips Collection, Washington
Right: Courtesy of the New York Public Library Picture Collection

40   Photograph by T. L. Gettings

41   Courtesy of the *Fort Worth Star-Telegram*

45   Original art by Gesine Ehlers for *Solar Age* magazine (cover, July 1976).

46   The Bettmann Archive

48   *Left:* From *The Collected Works of C. G. Jung*, translated by R. F. C. Hull, Bollingen Series XX. Vol. 5 *Symbols of Transformation*, copyright © 1956 by Princeton University Press. Germanic Sun Idol From the *Sachsisch Chronicon*, 1596. Fig. 4, p. 96. Reprinted by permission of Princeton University Press.
Right: Courtesy of American Heritage Publishing Co., Inc., *Worlds Around the Sun*, copyright © 1969

50   Illustration by Mark Schultz

CREDITS

CREDITS

page

100 The Phillips Collection, Washington

101 Courtesy of Flint Institute of Arts. Gift of Viola F. Bray Charitable Trust.

103 Illustration by Mark Schultz

104 Drawn by Mark Schultz after illustration by Henri A. Fluchere. Reproduced by permission of Harcourt Brace Jovanovich, Inc. from *Man and the Living World* by Karl von Frisch. Copyright 1949 by Deutscher Verlag, Berlin. Also by permission of Verlag Ullstein.

105 Drawn by Mark Schultz after illustration by Elizabeth Hollett Smith; from *The Living Clocks* by Richie R. Ward, illustrated by Elizabeth Hollett Smith. Copyright ©1971 by Richie R. Ward. Reprinted by permission of Alfred A. Knopf, Inc.; also by permission of William Collins Sons and Co., Ltd.

106 Collection: National Museum of Vincent van Gogh, Amsterdam

107 *Left:* Reproduced from *The New Book of Knowledge* by permission of the publisher, Grolier, Incorporated
Right: Illustration by Mark Schultz

108 Illustration by Mark Schultz

113 Illustration by John F. Carafoli

115 The Bettmann Archive

117, 118 Photograph by William Koechling. All rights reserved.

119 *Left:* Courtesy of Bausch & Lomb
Right: Illustration by Mark Schultz

121 © The Philosophical Research Society. Reprinted by Permission.

122 Illustrations by Mark Schultz

124 Aero Service Division of Western Geophysical Company of America

126 The Whitworth Art Gallery, University of Manchester

127 Permission of The Fine Arts Museums of San Francisco. Gift of Mr. Peter F. Young.

129 Photograph by William Koechling. All rights reserved.

page

131 Drawing by Ross: ©1978 The New Yorker Magazine, Inc.

133 Courtesy of the U.S. Department of Energy

134 Emanuel Hoffmann Foundation, Kunstmuseum, Basel

135 From *Color: A Survey in Words and Pictures, from Ancient Mysticism to Modern Science* by Faber Birren. Copyright ©1962, University Books.

137 Musée National d'Art Moderne, Centre Georges Pompidou, Paris, Copyright ©Estate of Max Ernst, 1980

139 Collections of the Library of Congress

140 Original art by Gesine Ehlers for *Solar Age* magazine (cover, August 1977)

141 Drawn by Mark Schultz after illustration by Anthony Colbert (originally appeared in the *Ecologist*, November 1977)

145 Illustration by Karen A. Schell

147 Courtesy of William A. Shurcliff

150 From *Design for a Limited Planet*, by Norma Skurka and Jon Naar. Copyright ©1976 by Norma Skurka and Jon Naar. Reprinted by permission of Ballantine Books, a Division of Random House, Inc. Photograph copyright ©1980 Jon Naar.

151 *Top left:* Courtesy of the U.S. Department of Energy
Bottom left: Courtesy of the American Yazaki Corporation
Right: Photograph by William Koechling. All rights reserved.

153 Smithsonian Institution Photo No. 33,573-A

154 Courtesy of the U.S. Department of Energy, photograph by Frank Hoffman

155 *Left:* Photograph by T. L. Gettings
Right: Photograph by William Koechling. All rights reserved.

156 Courtesy of the U.S. Department of Energy

157 Courtesy of the New York Public Library Picture Collection

page

158 Original art by Gesine Ehlers for *Solar Age* magazine (cover, April 1977)

159 Courtesy of the U.S. Department of Energy

160 *Left:* Courtesy of the New York Public Library Picture Collection
Right: Courtesy of the National Aeronautics and Space Administration

162 Courtesy of the American Yazaki Corporation

164 Smithsonian Institution Photo No. 48,281A

165 Courtesy of the New York Public Library Picture Collection

166 Illustrations courtesy of the U.S. Patent Office

167 Illustrations courtesy of the U.S. Patent Office

169 *Right:* Courtesy of the U.S. Department of Energy

170 Courtesy of the Trustees of Sir John Soane's Museum

171 *Top left:* Diagram from *Energy for Survival* by Wilson Clark with research by David Howell, illustrated by James K. Page, Jr. Copyright ©1974 by Wilson Clark. Reproduced by permission of Doubleday and Company, Inc.
Bottom left: Reprinted from *A Golden Thread: 2500 Years of Solar Architecture and Technology*, copyright 1980 by Ken Butti and John Perlin.
Right: Diagram by Jeffrey Cook from *Solar Architecture*, Gregory E. Franta and Kenneth R. Olson, copyright 1978 Ann Arbor Science Publishers, Inc., Ann Arbor, Michigan.

172 *Left:* From *New Scientist* magazine, July 13, 1978. Reprinted by permission of New Science Publications.
Right: Smithsonian Institution Photo No. 56,119

175 *Top:* Photograph by Carl Doney
Bottom: Photograph by T. L. Gettings

176 Photograph by T. L. Gettings

178 *Right:* Courtesy of the U.S. Department of Energy

179 Photograph by Carl Doney

page

180 From *Design for a Limited Planet,* by Norma Skurka and Jon Naar. Copyright ©1976 by Norma Skurka and Jon Naar. Reprinted by permission of Ballantine Books, a Division of Random House, Inc. Photograph copyright ©1980 Jon Naar.

181 Photograph by Richard Garrett

182 *Left:* Courtesy of the U.S. Department of Energy. Photograph by Frank Hoffman
Right: Courtesy of the U.S. Department of Energy. Photograph by Jack Schneider.

185 Courtesy of the U.S. Department of Energy. Photograph by the National Aeronautics and Space Administration.

189 Illustration by Karen A. Schell

191, 192 Courtesy of the U.S. Department of Transportation, Federal Highway Administration

193 Photograph by John Hoke

194 *Left:* Courtesy of the U.S. Department of Agriculture
Right: Courtesy of the U.S. Department of Energy

195 Photographs courtesy of the U.S. Department of Agriculture

196 LLL photograph for the U.S. Department of Energy

197 Courtesy of the Arizona Department of Transportation

199 Courtesy of the New Alchemy Institute

200 Courtesy of Mike Peters, *Dayton Daily News*

201 Courtesy of the New Alchemy Institute

205 Drawn by Mark Schultz after illustration in *Signs and Symbols of the Sun* by Elizabeth Helfman, © 1974. Houghton Mifflin/Clarion Books, New York

207 Illustration by Mark Schultz

209 Courtesy of the U.S. Department of Energy. Photograph by Jack Schneider.

210 Courtesy of the U.S. Department of Energy. Photograph by Dick Pebody.

Sidebar Credits

CREDITS

page

©1966 by Richard E. Byrd. Reprinted by permission of G. P. Putnam's Sons, New York.

140, 141 Specified excerpt from page 139 in *Four Seasons North* by Billie Wright. Copyright ©1973 by Billie Wright. Reprinted by permission of Harper & Row, Publishers, Inc.

142 From *The Politics of Energy* by Barry Commoner. Copyright © 1979 (New York: Alfred A. Knopf, Inc.). Reprinted by permission of the publisher.

149 Specified excerpt from page 86 in "Towards a Symbiotic Architecture" by Sean Wellesley-Miller from *Earth's Answer: Explorations of Planetary Culture at the Lindisfarne Conferences*, edited by Michael Katz, William P. Marsh, and Gail Gordon Thompson. Copyright ©1977 by the Lindisfarne Association. Reprinted by permission of Harper & Row, Publishers, Inc.

150 From *Design for a Limited Planet*, by Norma Skurka and Jon Naar. Copyright ©1976 by Norma Skurka and Jon Naar. Reprinted by permission of Ballantine Books, a Division of Random House, Inc.

151 From *Soft Energy Paths: Toward a Durable Peace* by Amory B. Lovins. Copyright ©1979. (New York: Harper Colophon Books). Reprinted by permission of the publisher.

152 From *Rays of Hope* by Denis Hayes. Copyright ©1977 by Worldwatch Institute. Reprinted by permission of Curtis Brown Ltd. and W. W. Norton Co., New York.

154 From *Solar Architecture* by Gregory E. Franta and Kenneth R. Olson, copyright ©1978 Ann Arbor Science Publishers, Inc., Ann Arbor, Michigan

159 From *The Ecological Conscience: Values for Survival* by Robert Disch. Copyright ©1970. (Englewood Cliffs, N.J.: Prentice Hall). Reprinted by permission of the publisher.

page

161 *Top:* From *Soft Energy Paths: Toward a Durable Peace* by Amory B. Lovins. Copyright © 1979. (New York: Harper Colophon Books). Reprinted by permission of the publisher.
Bottom: From *The Collected Poems of Wallace Stevens* by Wallace Stevens. Copyright ©1954. (New York: Alfred A. Knopf, Inc.). Reprinted by permission of the publisher.

168 Robert Williams, editor, *Toward a Solar Civilization*, MIT Press, ©1978 by the Massachusetts Institute of Technology

173 From *New Inventions in Low-Cost Solar Heating: One Hundred Daring Schemes Tried and Untried* by William A. Shurcliff. Copyright ©1979. Andover, Mass.: Brick House Publishing Co. Reprinted by permission of the publisher.

176, 177 From *Solar Age* magazine (June 1979, page 4). Reprinted by permission of the publisher.

186 From the film, *The Power to Change*, Third Eye Films, Cambridge, Mass.

204 From *Design for a Limited Planet* by Norma Skurka and Jon Naar. Copyright ©1976 by Norma Skurka and Jon Naar. Reprinted by permission of Ballantine Books, a Division of Random House, Inc.

214 From *The Collected Poems of Wallace Stevens* by Wallace Stevens. Copyright ©1954. (New York: Alfred A. Knopf, Inc.). Reprinted by permission of the publisher.

226, 227 From "An Optimal Solar Strategy" by Steven Nadis in *Environment* magazine (November 1979). Reprinted by permission of the author.

CREDITS

Scientific information in the materials accompanying the text was summarized from many sources, notably John Eddy, *A New Sun*, Washington, D.C.: NASA, 1979; Hoimar von Ditfurth, *Children of the Universe*, New York: Atheneum, 1976; Donald Menzell, *Our Sun*, Cambridge: Harvard University Press, 1959; and Patrick Moore, *The Sun*, New York: W. W. Norton, 1968. Myths were summarized from W. T. Olcott, *Sun Lore of All Ages*, New York: G. P. Putnam's Sons, 1914; Joseph Campbell, *The Masks of God: Oriental Mythology*, New York: Viking Press, 1962; Elisabeth Helfman, *Signs and Symbols of the Sun*, New York: Seabury Press, 1974; and Natalia Belting, *The Stars Are Silver Reindeer*, New York: Holt, Rinehart, Winston, 1966. Material on light and color was summarized from Faber Birren, *Color: A Survey in Words and Pictures, from Ancient Mysticism to Modern Science*, Secaucus, N.J.: University Books, 1962; J. Gordon Cook, *We Live by the Sun*, New York: Dial Press, 1957; Hyman Ruchlis, *The Wonder of Light: A Picture Story of How and Why We See*, New York: Harper and Row, 1960. Description of the Hopi Indian sun-greeting ritual is summarized from Frank Waters, *Pumpkin Seed Point*, Chicago: Swallow Press, 1969. Description of Arctic and tropic seasons was summarized from Anthony Smith, *The Seasons*, New York: Harcourt Brace Jovanovich, 1970.

INDEX

A

Abbott, Charles Greeley, 153
Adonis, as sun god, 58
African sun traditions, 58
Akhenaten, 54
Alaskan sun traditions, 58
Albacore, migrations of, 91
Alcohol distillation, solar, 181
Algae, marine, biological clock of, 104
Allen, Marcus, 47
Amaterasu, as sun god, 50
American Indian sun traditions, 55–62, 65
Anxiety, caused by limits, 27
Apollo, as sun god, 17, 49, 58
Ark, of New Alchemy Institute, 198–206
Arthritis, sunlight and, 128
Asabinus, as sun god, 58
Aspirin, effect on body, 116
Astrological charts, 47
Astrology for the New Age (Allen), 47
Aton, as sun god, 54
Atum, as sun god, 58
Aurora borealis, 133
Automobile industry, 193–95
Aztec sun traditions, 58
Aztec temples, 53

B

Baal, as sun god, 58
Bacchus, as sun god, 58
Barnhart, Earle, 200
Bass, striped, migration of, 91
Behavior, personal, affected by weather, 115–18
Belenus, as sun god, 58
Bel-Samen, as sun god, 58
Bergmark, David, 202

Biological clocks, of animals, 104
Biomass, importance of, 96, 109, 209
Bluefish, migrations of, 91
Bolen, Joseph, 137
Bolivian Indian sun traditions, 138
Brecht, Bertolt, 47
Brookhaven National Laboratory, 34
Buckingham, Jerry, 181
Buddha, as sun god, 52
Byron, Lord, 85

C

Calder, Nigel, 85
Cancer, sunlight and, 129–30, 132
Carbon dioxide, in the atmosphere, 84, 108
Cell regeneration, ultraviolet rays and, 124
Chambers, Don, 181
Cherokee sun traditions, 55, 57
Chinese sun traditions, 58
Christian sun traditions, 66
Chromosphere, of sun, 20
Chrysler, Walter, 166
Circadian rhythm, 104
Cities, loss of natural presences in, 98
Cockroaches, biological clocks of, 107
Conservation, of energy, 205–6
Copernicus, 3, 12
Core, of sun, 20
Cornerstones building school, 148
Cornman, Ron, 174
Corona, of sun, 20, 30, 102
Cosmic calendar, of Carl Sagan, 15–17
Cosmic rays, from sun, 84
Crab, fiddler, biological clock of, 104, 107
Crane, Siberian, migrations of, 92
Crombie, Duane, 181
Cronus, as sun god, 58

D

Daphne, as sun goddess, 70
Daylength, importance of, 104
Dragons of Eden, The (Sagan), 15
Dust, in the atmosphere, 84, 85

E

Easter Island, monoliths on, 58
Eclipse, solar, 30, 31
Eddy, Jack, 32–36
Egyptian sun myth, 52, 58
Electric fence, solar-powered, 183
Electricity, from the sun, 184, 198
Equinoxes, precession of, 83
Eskimo sun myth, 52, 58
European sun traditions, 58

F

Farrell, Thomas, 110
Faventinus, 11
Fish, migrations of, 91
Fish farming, 203
Flowers, biological clocks of, 105
Fluorescent light, health and, 129–31
Ford, Henry, 166
Fossil fuels, limits of, 20
Frankenstein (Shelley), 85
Fraunhofer, Joseph von, 32
Fregosi, Claudia, 49
French sun traditions, 58
Fuller, R. Buckminster, 159

G

Galileo, 14, 38, 45
Gamma rays, from sun, 120
Gom, as sun god, 58

Greek sun traditions, 58
Greenhouse, solar, 177–81
Greenhouse effect, 29, 84, 108

H

Halos, around sun, 102
Hammarlund, Ole, 200
Hammon, as sun god, 58
Harrison, Benjamin, 55
Hawaiian sun traditions, 72
Hawks, migrations of, 91
Health, sunlight and, 115–43 passim
Health and Light (Ott), 125
Hercules, as sun god, 58
Hindu sun traditions, 58
Hopi Indian sun traditions, 57, 58, 127, 138
Houses, solar, 147–87 passim
Hu, as sun god, 58
Humidity, relative, 103

I

Ice Age, 81–83
Ice-out, in spring, 77–80
Inca sun traditions, 58
Indru, as sun god, 58
Infrared rays, from sun, 120
Ionosphere, 109
Irish sun traditions, 58
Izanagi, 50

J

Janus, as sun god, 58
Japanese sun myth, 50, 58
Jet lag, 105
Jet streams, 84–85
Jonathan Livingston Seagull (Bach), 47

Joss, Paul C., 36–37, 38
Jupiter, as sun god, 58

K

Katz, Arthur, 157
Kinsey, Alfred, 116

L

Light, physical characteristics of, 125
Lisa, as sun god, 54
Lord of the Rings, The (Tolkien), 47
Lucretius, 9

M

Mainge, Hilde, 200
Mandoo, as sun god, 58
Maui, as sun god, 58
Mawu, as sun god, 54
Mayan sun traditions, 58
Melanin, rickets and, 124
Mercury, as sun god, 58
Mesopotamian sun traditions, 45
Methane digestors, 209
Mexican sun myth, 52
Midsummer Day, celebration of, 68
Midsummer Night's Dream, A (Shakespeare), 70
Migrations, animal, 91–93
Mithras, as sun god, 58
MITRE Corporation, 213
Moloch, as sun god, 58
Moni, as sun god, 58
Mussorgsky, Modest, 70
Myth, 45–73 passim
    need for, 19
Myths of the Sun (Olcott), 52

N

Nearing, Scott and Helen, 180
Neer, Robert, 134–35
Neutrinos, 34, 38
New Alchemy Institute, 197–206
Newman, Daniel and Sandra, 180
New Zealand sun traditions, 58
Nexhequiriac, as sun god, 52
Night on Bald Mountain (Mussorgsky), 70
Northrup, Lynn, 162
Nuclear energy, 4, 206–7

O

Olcott, William Tyler, 52
Osiris, as sun god, 54, 58
Ott, John N., 125–28
Ozone layer, 109–11

P

Pan, as sun god, 58
Pearson, Bill, 184
Photosphere, of sun, 19, 20
Photovoltaic cells, 184
Pineal gland, sunlight and, 128–29
Plages, of sun, 20
Plants, use of solar energy by, 96
Plato, 50
Plover, migrations of, 91
Pluto, as sun god, 58
Polar bears, migrations of, 92
Polar ice caps, solar eruptions and, 84
Polynesian sun traditions, 58
Ponds, solar, 203
Prigogine, Ilya, 9
Prisms, 36
Ptah, as sun god, 58

Ptolemy, 12
Precession of equinoxes, 83
Pyramids, of Incas, 58

R

Ra, as sun god, 58
Radiative interior, of sun, 20
Radio reception
    ionosphere and, 109
    solar eruptions and, 84
Rainbow
    colors of, 118–20
    myths about, 121
Republic, The (Plato), 50
Rickets, sunlight and, 122–23
Roman sun traditions, 58

S

Sagan, Carl, 15
Saint Peter's Cathedral, orientation of, 65
Sargent, Francis, 191
Saturn, as sun god, 58
Scandinavian sun traditions, 58
Scheiner, Christopher, 23
Scholl, Lisette, 117
Science, volatility of, 38–40
Seasons, getting reacquainted with, 86–90
Second Law of Thermodynamics, 9
Serapis, as sun god, 58
Set, as sun god, 58
Sex, of sun, in myths, 55
Shakespeare, 68
Shamash, as sun god, 51, 58
Shapiro, Irwin, 37
Shelley, Mary, 85
Shoshoni Indian sun traditions, 59

Shurcliff, William A., 147
Siva, as sun god, 58
Skylab, 33
Smith, Kendric C., 138
Solar Energy Research Institute, 149
Solar flares, 20, 84
Solar prominences, 20
Solar Survival, 203
Solar village, prototype of, 29–30, 40
Solar wind, 20, 28
Solar year, true, 88–89
Solsearch Architects, 200
Solstices, celebration of, 65, 68–70
Spectroscope, 32, 33
Stonehenge, 48, 58, 70
Sumerian sun traditions, 45
Summer solstice, celebration of, 68–70
Sun
        nuclear processes in, 18
        physical characteristics of, 15, 19, 21
        shrinkage of, 32–36
Sun Grumble (Fregosi), 49
Sunset and rise, weather predictions from, 99
Sunspaces, on houses, 177–81
Sunspots, 20, 24, 83–84
Surya, as sun god, 63
Swallow, European, migrations of, 92
Swan, Bewick's, migrations of, 92

T

Tambora volcano, 85
Technological society, disadvantages of, 19
Tecumseh, 55
Television reception, solar eruptions and, 84
Thoreau, Henry, 116
Todd, John, 198

Toltec sun traditions, 58
Trees, deciduous, seasonal changes of, 93

U

Ultraviolet rays, 120–22, 124, 132
United States Navy, use of full-spectrum lights
    by, 136
Urotal, as sun god, 58
Ute Indian sun traditions, 59

V

Violence, in American cities, 98
Viracocha, as sun god, 58
Vishnu, as sun god, 58
Visionetics (Scholl), 117
Vitzliputzli, as sun god, 58
Volcanic eruptions, weather changes from, 85
Vulcan, as sun god, 58

W

Weather changes
    caused by solar eruptions, 84
    of coming century, 85–86
    sun's angle and, 102
Weather Machine, The (Calder), 85
Wells, Malcolm, 186
Whales, humpback, migrations of, 92
Wilson, Charles, 193
Winds, caused by sun, 99–102
Woese, C. R., 126
Wolleston, William, 32
Wurtman, Richard J., 129

X

X-rays, from sun, 120